Artificial Intelligence, Big Data, Algorithms and Industry 4.0 in Firms and Clusters

This volume offers a wide-ranging discussion on the interrelations among AI, algorithms, big data, and Industry 4.0 to understand the importance of these new paradigms for the development of firms, districts, clusters, cities, regions, and innovation.

Drawing on theoretical, empirical, and qualitative studies and using local perspectives, the chapters in this book explore theoretical aspects of AI and its evolution in social sciences, focusing on Industry 4.0, smart cities, big data, and other related topics. They examine the role of industrial robots in employment, productivity, and knowledge absorption in industrial districts. They also discuss innovation in the context of local production systems, AI ecosystems, and the growth and potential of the metaverse. Taken together, the book offers insights to help understand the new dynamics generated by the advent of these technologies and how they may affect regions, cities, clusters, industries, and organizations, and identifies avenues for future research in the development of new trajectories for clusters and firms.

This book will be a key resource for scholars and advanced students in the fields of economics, geography, architecture, planning, and management as well as for interdisciplinary researchers who want to learn more about the development of new technologies, the relevance of AI, big data and Industry 4.0 for firms and in relation to their adoption in clusters. This book was originally published as a special issue of *European Planning Studies*.

Luciana Lazzeretti is Full Professor of Management at the Department of Economics and Management, University of Florence, Italy.

Rafael Boix-Domenech is Full Professor of Economic Structure at the Faculty of Economics, Department of Economic Structure, University of Valencia, Spain.

Jose-Luis Hervas-Oliver is Full Professor of Management at the Departamento de Organización de Empresas, Universitat Politècnica de València, Spain.

Niccolò Innocenti is Assistant Professor of Management at the Department of Economics and Management, University of Florence, Italy.

Artificial Intelligence, Big Data, Algorithms and Industry 4.0 in Firms and Clusters

Edited by

Luciana Lazzeretti, Rafael Boix-Domenech, Jose-Luis Hervas-Oliver, and Niccolò Innocenti

Routledge
Taylor & Francis Group
LONDON AND NEW YORK

First published 2025
by Routledge
4 Park Square, Milton Park, Abingdon, Oxon OX14 4RN

and by Routledge
605 Third Avenue, New York, NY 10158

Routledge is an imprint of the Taylor & Francis Group, an informa business

British Library Cataloguing in Publication Data
A catalogue record for this book is available from the British Library

ISBN13: 978-1-032-84693-4 (hbk)
ISBN13: 978-1-032-84853-2 (pbk)
ISBN13: 978-1-003-51599-9 (ebk)

DOI: 10.4324/9781003515999

Typeset in Minion Pro
by Newgen Publishing UK

Publisher's Note
The publisher accepts responsibility for any inconsistencies that may have arisen during the conversion of this book from journal articles to book chapters, namely the inclusion of journal terminology.

Contents

Citation Information

The chapters in this book were originally published in the *European Planning Studies*, volume 31, issue 7 (2023). When citing this material, please use the original page numbering for each article, as follows:

Introduction

Artificial intelligence, big data, algorithms and Industry 4.0 in firms and clusters
Luciana Lazzeretti, Rafael Boix-Domenech, Jose-Luis Hervas-Oliver and Niccolò Innocenti
European Planning Studies, volume 31, issue 7 (2023), pp. 1297–1303

Chapter 1

The emergence of artificial intelligence in the regional sciences: a literature review
Luciana Lazzeretti, Niccolò Innocenti, Martina Nannelli and Stefania Oliva
European Planning Studies, volume 31, issue 7 (2023), pp. 1304–1324

Chapter 2

Artificial intelligence in hospitality and tourism. State of the art and future research avenues
Martina Nannelli, Francesco Capone and Luciana Lazzeretti
European Planning Studies, volume 31, issue 7 (2023), pp. 1325–1344

Chapter 3

Learning as imitation or mimesis: how 'smart' is machine learning for its planning controllers?
Philip Cooke
European Planning Studies, volume 31, issue 7 (2023), pp. 1345–1357

Chapter 4

Does regional innovation policy really work for Industry 4.0? Evidence for industrial districts
Jose-Luis Hervas-Oliver, Sofia Estelles-Miguel, Marta Peris-Ortiz and Jose Antonio Belso-Martínez
European Planning Studies, volume 31, issue 7 (2023), pp. 1358–1376

Chapter 5

The actors' role in industrial districts facing the challenge of technologies 4.0 between absorption and dissemination
Niccolò Fiorini, Tommaso Pucci, Elena Casprini and Lorenzo Zanni
European Planning Studies, volume 31, issue 7 (2023), pp. 1377–1396

Chapter 6

Robots and the productivity of local manufacturing systems in Emilia-Romagna: the mediating role of occupational similarity and complexity
Roberto Antonietti, Luca Cattani and Giulio Pedrini
European Planning Studies, volume 31, issue 7 (2023), pp. 1397–1421

Chapter 7

Living innovation machines: modelling innovation in time and space variable-geometry territorial units using machine learning
Rafael Boix-Domenech, Vittorio Galletto, Fabio Sforzi and Francesco Capone
European Planning Studies, volume 31, issue 7 (2023), pp. 1422–1442

Chapter 8

The mechanisms underlying the emergence of innovation ecosystems: the case of the AI ecosystem in Montreal
Nasrin Sultana, Ekaterina Turkina and Patrick Cohendet
European Planning Studies, volume 31, issue 7 (2023), pp. 1443–1465

Chapter 9

How real will the metaverse be? Exploring the spatial impact of virtual worlds
Pier Luigi Parcu, Maria Alessandra Rossi, Niccolò Innocenti and Chiara Carrozza
European Planning Studies, volume 31, issue 7 (2023), pp. 1466–1488

Notes on Contributors

Roberto Antonietti, "Marco Fanno" Department of Economics and Management, University of Padova, Italy.

Jose Antonio Belso-Martínez, Universidad Miguel Hernández, Elche, Spain.

Rafael Boix-Domenech, Faculty of Economics, Department of Economic Structure, University of Valencia, Valencia, Spain.

Francesco Capone, Department of Economics and Management, University of Florence, Firenze, Italy.

Chiara Carrozza, European University Institute, Robert Schuman Centre for Advanced Studies, Centre for a Digital Society, Florence, Italy.

Elena Casprini, Department of Business and Law, University of Siena, Italy.

Luca Cattani, Gran Sasso Science Institute (GSSI), L'Aquila, Italy.

Patrick Cohendet, Department of International Business, HEC Montreal, Canada.

Philip Cooke, Mohn Institute for Innovation and Regional Development, Western Norway University of Applied Sciences, Bergen, Norway.

Sofia Estelles-Miguel, Universitat Politècnica de València, Valencia, Spain.

Niccolò Fiorini, Department of Business and Law, University of Siena, Italy.

Vittorio Galletto, Institut d'Estudis Regionals i Metropolitans de Barcelona, Universitat Autònoma de Barcelona, Barcelona, Spain.

Jose-Luis Hervas-Oliver, Departamento de Organización de Empresas, Universitat Politècnica de València, Valencia, Spain.

Niccolò Innocenti, Department of Economics and Management, University of Florence, Italy.

Luciana Lazzeretti, Department of Economics and Management, University of Florence, Italy.

Martina Nannelli, Department of Economics and Management, University of Florence, Italy.

Stefania Oliva, Department of Economics and Management, University of Florence, Italy.

Pier Luigi Parcu, Robert Schuman Centre for Advanced Studies, Centre for a Digital Society, European University Institute, Florence, Italy.

Giulio Pedrini, Faculty of Economics and Law, Kore University of Enna, Italy.

Marta Peris-Ortiz, Universitat Politècnica de València, Valencia, Spain.

Tommaso Pucci, Department of Business and Law, University of Siena, Italy.

Maria Alessandra Rossi, Robert Schuman Centre for Advanced Studies, Centre for a Digital Society, European University Institute, Florence, Italy; Department of Economics, G. D'Annunzio University of Chieti-Pescara, Pescara, Italy.

Fabio Sforzi, Department of Economics and Management, University of Parma, Italy.

Nasrin Sultana, Department of International Business, HEC Montreal, Canada.

Ekaterina Turkina, Department of Entrepreneurship and Innovation, HEC Montreal, Canada.

Lorenzo Zanni, Department of Business and Law, University of Siena, Italy.

Introduction: artificial intelligence, big data, algorithms and Industry 4.0 in firms and clusters

Luciana Lazzeretti, Rafael Boix-Domenech, Jose-Luis Hervas-Oliver and Niccolò Innocenti

ABSTRACT

This collection on 'Artificial intelligence, big data, algorithms and Industry 4.0 in firms and clusters' is introduced exploring the themes discussed by the nine papers and grouped into three categories to uncover new dynamics and identify future research opportunities for clusters and organizations in these transformative times. The first group explores theoretical aspects of AI and its evolution in social sciences, focusing on industry 4.0, smart cities, big data, and other related topics. The second group examines the role of industrial robots in employment, productivity, and knowledge absorption in industrial districts. The third group discusses innovation in the context of local production systems, AI ecosystems, and the growth and potential of the Metaverse.

1. Introduction

The last years have seen a strong increase in interest and studies on artificial intelligence (AI) and robots and their potential to trigger a new industrial revolution. Several articles have been written to explore its paradigm fields of application and impact on research areas, while others have applied computational artificial intelligence tools such as machine learning, deep learning and neural networks to study specific industries or sectors (Furman and Seamans 2019, Floridi and Cowls 2019; Acemoglu and Restrepo 2020). While the disruptiveness of AI is expanding exponentially in all fields of the economy, there are still some underdeveloped fields regarding the studies, particularly the studies regarding the role of AI for regional sciences and clusters seem to be under-developed (Lazzeretti 2023). In addition, little retrospective on this evolution has been made so far and thus, we still know little about the impact that this new paradigm is making in regional sciences and particularly looking at clusters and the firm level.

Artificial Intelligence (AI) refers to advanced technologies that exhibit human-like intelligence, including machine learning or autonomous robotics and vehicles, among others (see Furman and Seamans 2019; Lazzeretti et al. 2022). Along with AI, Industry 4.0 and all related digital technologies, such as the Internet of Things (IoT), 3D printing, Cloud Computing or Cyber-Physical systems, among others, are challenging existing business models, products and industries, offering new business opportunities and also altering existing value propositions (e.g. Porter and Heppelmann 2014, 2015). These enabling technologies are evidenced to improve firm performance, maximize efficiency, and decrease set-up costs, errors and machine downtime. Also, they enhance the quality of the products and the reduction of production waste, while increasing production capacity (see Büchi, Cugno, and Castagnoli 2020; Dalenogare et al. 2018).

Industrial and innovation policies are supporting digitization of the economy. The European Union, like many countries such as UK or USA, is facilitating the adoption of digital technologies in businesses, industries and even in regions (see the European Expert Group on Clusters 2021). A new paradigm for understanding the new phenomenon and its related policies and plans is ascribed to the Digital Single Market initiative by the European Commission,[1] concomitant with many national policy initiatives to stimulate the digital shift. At the regional level, The European Expert Group on Clusters (2021) points out the importance of clusters to the green and digital transitions. An example of the regional approach followed by the EU in the Smart Specialization Strategy (RIS3) is the initiative of the Digital Innovation Hubs,[2] targeting regions and allowing local businesses and organizations to follow a bottom-up process for setting up digital platforms for increasing Industry 4.0, especially targeting SMEs (see more in Hervas-Oliver et al. 2021).

Zooming into the regional setting, work is starting to audit Industry 4.0 in regions and districts/clusters (e.g. Bianchi, Durán, and Labory 2019; De Propris and Bailey 2020; Bettiol, Di Maria, and Micelli 2020; Hervás-Oliver 2021; Hervas-Oliver 2022; Lazzeretti 2022; Balland and Boschma 2021; Doloreux and Turkina 2021), recognizing Artificial Intelligence as an emergent line of inquiry crossing many works of literature and positioning them in the regional realm (Lazzeretti et al. 2022).

In industrial districts and clusters, the literature has pointed out that collective initiatives, based on cooperation and open doors, might support the digitization of territories, by co-creating place-based innovation policies to digitize specific territories (Hervas-Oliver et al. 2021; Pagano et al. 2021). Cooperation and competition, therefore, constitute a powerful leverage to digitize districts and their SMEs.

Industry 4.0 and related technologies are also applied to SMEs, showing how business models are reshaped (Müller, Buliga, and Voigt 2018) and how these firms present less probability to adopt these new technologies due to their scarcity of resources (e.g. Mittal et al. 2018) their reliance on local networks (Hervas-Oliver 2022) and external sources of knowledge (Petruzzelli et al. 2022).

All in all, different research questions are starting to be considered, such as how does Industry 4.0 shape industrial clusters and regions? What policy initiatives foster digitization in clusters and regions? What advantages or disadvantages have clustered SMEs for Industry 4.0? Other questions are still unanswered and this special issue brings them to this emerging conversation within the realm of the geography of innovation literature.

Given these considerations, the Special Issue wants to enlarge this debate to offer a wide discussion on the interrelation among AI, algorithms, big data and Industry 4.0 using a local perspective to understand the importance of these new paradigms for the development of firms, districts, clusters, city, regions and innovation (Cooke 2017;

Götz and Jankowska 2017; Cooke 2019; Cséfalvay 2020; Zukin 2020; Scherrer 2021; Lazzeretti 2022).

In particular, the special issue has a twofold objective. Firstly, understanding the new dynamics generated by the advent of these technologies and how they may affect regions, cities, clusters, industries and organizations. Secondly, identifying avenues for future research in the development of new trajectories for clusters and firms in these challenging times.

2. Contents of the special issue

The present special issue departs from a solid base of papers that were presented during the third and fourth editions of the Rethinking Cluster Conference that were held in Valencia and Florence in the 2020 and 2021 respectively. This special issue is composed of nine papers that were strongly connected with the themes of 'Artificial Intelligence, Big data, Algorithms and Industry 4.0 in firms and clusters' in the actual challenging times that see the increasing importance of digital transformation. However, throughout the special issue, we may distinguish several different perspectives, approaches and trajectories of the papers that compose the collection.

It is possible to identify a first group, composed of three papers that cover theoretical aspects of the issues related to the evolution of the studies of artificial intelligence in social sciences and the role of learning and machine learning.

The first paper by Lazzeretti et al. (2022) represents a deep reflection and a theoretical discussion regarding the artificial intelligence topic in social sciences and more deeply in regional sciences. After a first discussion on the definition of AI, it develops a bibliometric analysis to follow the evolution of this literature over 1986–2020 identifying the main authors, journals and topics covered by this literature. Then, developing an in-depth review of the most recent papers, it identifies five main groups of topics, the one related to 'Industry 4.0', a group related to 'Smart cities', those connected with 'Big data and social media', obviously an entire group related to 'Artificial intelligence' and finally a group including the remaining technologies related to AI 'Robotization, IoT, augmented and virtual reality'.

The second paper, authored by Nannelli, Capone, and Lazzeretti (2023) deals with the recent strong transformation brought by information and communication technologies (ICT) and AI technologies to the tourism and hospitality industry. The article, based on papers published until 2021, firstly develops a bibliometric analysis also using network analysis to map the intellectual structure of the AI tools in tourism and then, based on the most relevant publications retrieved, develops a critical literature review to identify the main themes and trends of the literature. The main research themes identified relate to 'Big data', 'Augmented reality', 'Virtual reality', 'Smart tourism' and 'Covid-19 pandemic'.

While the last paper of this first group by Cooke (2022) explores a new idea regarding the role of 'learning' and its evolution in the last decades, relating this evolution also with the recent introduction of new information technology and machine learning. The development of the paper leads to some relevant conclusions, regarding the 'knowledge economy' of the 1980s, which caused concern that 'lifelong learning' was needed to cope with the economic change and the following neoliberal politics avoiding planning and learning. And the most recent digital economy and 'spontaneous' markets as the sole origin of rational change show many problems and dark sides. It suggested that it is important to review and re-regulate the 'learning' arena.

Then the second group of papers focus on Industry 4.0 and the role of industrial robots in employment and productivity. The first paper, authored by Hervas-Oliver et al. (2022) presents a qualitative analysis based on 24 interviews and a review of the literature on the adoption of I.4.0 in three different Marshallian Industrial Districts (MID; one on footwear, one on toys and one on ceramic tiles) of the Valencian region in Spain. In this article, the purpose is to compare different innovation policies aimed at facilitating awareness and the adoption of I.4.0 in those three MIDs considered. The results clearly show that one-size-fits-all Industry 4.0 policies can't be identified for all the different MIDs. Rather, what works better are collective actors leading place-based collective actions that are bottom-up and co-designed with public and private local actors.

The second paper, by Fiorini et al. (2022), deals with the effect of I.4.0 in industrial districts, showing that industrial districts may be a relevant field for the empirical investigation to understand how the knowledge linked to the I.4.0 technological paradigm can be absorbed and disseminated. The analysis is based on more than 100 interviews with firms based in four Tuscan IDs: Prato (textile), Arezzo (jewellery), Florence (leather) and Santa Croce (tanning). The interviews gathered information on the two main dimensions considered (absorption and dissemination) and identified different behaviours of the four considered IDs. The results show what can lead to District 4.0 is not the size or the technologies adoption alone, but the presence of competencies 4.0, culture 4.0 and the absorption and dissemination through collaboration.

The paper that closes this group is authored by Antonietti, Cattani, and Pedrini (2023). The paper departs from the growing literature that connects industrial robots with employment and productivity following the worrying idea that robots could substitute a part of human jobs shortly. The analysis is based on one of the main manufacturing regions in Europe (Emilia-Romagna) and encompasses 15 manufacturing industries and 39 local labour systems from 2008 to 2017. The results show that higher robots adoption led to a decline in high routine employment while it does not increase labour productivity or manufacturing value-added. However, the negative effect of robotization on employment decreases in the presence of high cross-industry occupational similarity and within industrial districts.

A third group of papers dealing with innovation and composed of both quantitative and qualitative contributions is provided by three papers. The first paper, authored by Boix-Domenech et al. (2022) is based on the idea that local communities are the real unit of generation of innovation and that they are constantly changing their boundaries and specializations, behaving like 'living innovation machines'. The authors use dynamic territorial units (local production systems that modify their geographical boundaries and specialization over time) and machine learning methods to explain the change in the innovative intensity of local production systems (LPSs) in Spain between 1991 and 2014. The results suggest that the type of LPS of origin continues to be relevant in explaining their innovation processes and the transformation to a different type of LPS does not necessarily implies changes or improvements in innovative intensity.

The second paper of this group by Sultana, Turkina, and Cohendet (2023) is based on the study of the AI ecosystem in Montreal, looking at the mechanisms that favour the emergence of innovation ecosystems. The study is based on the abductive approach, which uses qualitative content analysis and social network analysis. The data used are partly from primary sources (semi-structured interviews) and partly from secondary sources (data on 19 key actors and 102 organizations). The results show a bottom-up approach in the emergence of the AI ecosystem in Montreal and show that the essential mechanism underlying the emergence of innovation ecosystems in Montreal is the articulation of a series of innovation by key actors.

Finally, the paper that closes the special issue was authored by Parcu et al. (2023) and deals with the growth of the expectations related to the advent of the Metaverse during the last years and which suddenly showed also a decline in attention, investments and financial performances. The aim of the paper is to go beyond the hype looking at several aspects of the Metaverse and investigating the firms, industries and the spatial effects that may represent the advent of the Metaverse in two possible scenarios. If the Metaverse will remain in its evolution as a fashion label, applications will be limited to replicating existing human activities in an immersive virtual environment and it will be a physical reality to drive and shape the Metaverse. While if the Metaverse will maintain its expectations enabling new forms of social interaction, and substantially altering key economic concepts and institutions we will have a Metaverse shaping reality view.

We hope the nine articles in this special issue will draw attention to these new technologies which may be considered potential breakthroughs and particularly when analyzed at the regional, city, cluster or firm level, leading to a new wave of innovation across a range of industries and also can stimulate further research. Following this, in concluding this introduction we aim to indicate some possible developments of the outlined debate. We firstly refer to the research agenda that arises from the papers that compose this issue and to the various risks and opportunities mentioned that concern the evolution of these new technologies that are already changing the competitive environment for companies and clusters and that are expected to bring additional challenges in the coming years (Lazzeretti 2023).

Of course, some of the results achieved cannot be generalized without due caution, as they are tested for certain countries, regions or specific technologies. For this reason, it appears clear that these topics need further analysis to confirm some of the achieved results for different areas or prove that the effects are context-specific. These create opportunities for additional research, and this is even more important considering the continuous and rapid evolution of these technologies. Think for instance of the Metaverse as discussed in the last paper of this selection and the uncertainty of its evolution, or the possibly disruptive effect of the use of AI recently evolved in an unexpectedly rapid use and adoption of ChatGPT, which will possibly lead to a revolution form an economic, social and regulatory perspective. Like the already mentioned I.4.0 and robotization, which, of course, will call for additional planning and policies aimed at dealing with the changes in the business models of firms and the industrialization of territories.

Notes

1. https://ec.europa.eu/info/strategy/priorities-2019-2024
2. https://digital-strategy.ec.europa.eu/en/activities/edihs

Disclosure statement

No potential conflict of interest was reported by the author(s).

Funding

Dr. Hervas-Oliver and Dr. Boix-Domenech are financially funded by PID2021-128878NB-100 (MCIN/AEI/10.13039/501100011033).

References

Acemoglu, D., and P. Restrepo. 2020. "The Wrong Kind of AI? Artificial Intelligence and the Future of Labour Demand." *Cambridge Journal of Regions, Economy and Society* 13 (1): 25–35. doi:10.1093/cjres/rsz022.

Antonietti, R., L. Cattani, and G. Pedrini. 2022. "Robots and the Productivity of Local Manufacturing Systems in Emilia-Romagna: The Mediating Role of Occupational Similarity and Complexity." *European Planning Studies*, 1–25. doi:10.1080/09654313.2022.2161816.

Balland, P. A., and Boschma, R. 2021. Mapping the potentials of regions in Europe to contribute to new knowledge production in Industry 4.0 technologies. *Regional Studies* 55 (10-11): 1652–1666.

Bettiol, M., E. Di Maria, and S. Micelli, eds. 2020. *Knowledge Management and Industry 4.0: New Paradigms for Value Creation*. Vol. 9. Heidelberg: Springer.

Bianchi, P., C. Durán, and S. Labory. 2019. *Transforming Industrial Policy for the Digital age: Production, Territories and Structural Change*. Cheltenham: Edward Elgar Publishing.

Boix-Domenech, R., V. Galletto, F. Sforzi, and F. Capone. 2022. "Living Innovation Machines: Modelling Innovation in Time and Space Variable-Geometry Territorial Units Using Machine Learning." *European Planning Studies*, 1–21. https://doi.org/10.1080/09654313.2022.2134727

Büchi, G., M. Cugno, and R. Castagnoli. 2020. "Smart Factory Performance and Industry 4.0." *Technological Forecasting and Social Change* 150: 119790. doi:10.1016/j.techfore.2019.119790.

Cooke, P. 2017. "Digital Tech'and the Public Sector: What New Role after Public Funding?" *European Planning Studies* 25 (5): 739–754. doi:10.1080/09654313.2017.1282067.

Cooke, P. 2019. "Responsible Research and Innovation? From FinTech's 'Flash Crash'at Cermak to Digitech's Willow Campus and Quayside." *European Planning Studies* 27 (12): 2376–2393. doi:10.1080/09654313.2018.1556610.

Cooke, P. 2022. "Learning as Imitation or Mimesis: How 'Smart' is Machine Learning for its Planning Controllers?" *European Planning Studies*, 1–13. doi:10.1080/09654313.2022.2124102.

Cséfalvay, Z. 2020. "Robotization in Central and Eastern Europe: Catching up or Dependence?" *European Planning Studies* 28 (8): 1534–1553. doi:10.1080/09654313.2019.1694647.

Dalenogare, L., G. Benitez, N. Ayala, and A. Frank. 2018. "The Expected Contribution of Industry 4.0 Technologies for Industrial Performance." *International Journal of Production Economics* 204: 383–394. doi:10.1016/j.ijpe.2018.08.019.

De Propris, L., and D. Bailey. 2020. *Industry 4.0 and Regional Transformations*. New York, NY: Taylor & Francis.

Doloreux, D., and Turkina, E. 2021. New path creation in the artificial intelligence industry: regional preconditions, new actors and their collective actions, and policies. *Regional Studies* 55 (10-11): 1751–1763.

European Expert Group on Clusters. 2021. *European Expert Group on Clusters Recommendation Report Directorate-General*. Bruxelles: European Commission.

Fiorini, N., T. Pucci, E. Casprini, and L. Zanni. 2022. "The Actors' Role in Industrial Districts Facing the Challenge of Technologies 4.0 Between Absorption and Dissemination." *European Planning Studies*, 1–20. doi:10.1080/09654313.2022.2161299.

Floridi, L., and J. Cowls. 2019. "A Unified Framework of Five Principles for AI in Society." *Harvard Data Science Review* 1 (1): 1–15.

Furman, J., and R. Seamans. 2019. "AI and the Economy." *Innovation Policy and the Economy* 19 (1): 161–191. doi:10.1086/699936.

Götz, M., and B. Jankowska. 2017. "Clusters and Industry 4.0–do They Fit Together?" *European Planning Studies* 25 (9): 1633–1653. doi:10.1080/09654313.2017.1327037.

Hervas-Oliver, J. 2022. "Industry 4.0 in Industrial District SMEs: Understanding Collective Knowledge Transfer by Research and Transfer Institutes." *Competitiveness Review: An International Business Journal* 31 (5): 647–666. doi:10.1108/CR-06-2022-0075.

Hervas-Oliver, J., S. Estelles-Miguel, M. Peris-Ortiz, and J. Belso-Martínez. 2022. "Does Regional Innovation Policy Really Work for Industry 4.0? Evidence for Industrial Districts." *European Planning Studies*, 1–19. doi:10.1080/09654313.2022.2135368.

Hervas-Oliver, J., G. Gonzalez-Alcaide, R. Rojas-Alvarado, and S. Monto-Mompo. 2021. "Emerging Regional Innovation Policies for Industry 4.0: Analyzing the Digital Innovation Hub Program in European Regions." *Competitiveness Review: An International Business Journal* 31 (1): 106–129. doi:10.1108/CR-12-2019-0159.

Hervás-Oliver, J. 2021. "Industry 4.0 in Industrial Districts: Regional Innovation Policy for the Toy Valley District in Spain." *Regional Studies* 55 (10-11): 1775–1786. doi:10.1080/00343404.2021.1939861.

Lazzeretti, L. 2022. "What is the Role of Culture Facing the Digital Revolution Challenge? Some Reflections for a Research Agenda." *European Planning Studies* 30 (9): 1617–1637. doi:10.1080/09654313.2020.1836133.

Lazzeretti, L. 2023. *The Rise of Algorithmic Society and the Strategic Role of Arts and Culture.* Cheltenham.: Edward Elgar.

Lazzeretti, L., N. Innocenti, M. Nannelli, and S. Oliva. 2022. "The Emergence of Artificial Intelligence in the Regional Sciences: A Literature Review." *European Planning Studies*, 1–21. doi:10.1080/09654313.2022.2101880.

Messeni Petruzzelli, A., Murgia, G., and Parmentola, A. 2022. How can open innovation support SMEs in the adoption of I4. 0 technologies? An empirical analysis. *R&D Management* 52 (4): 615–632.

Mittal, S., Khan, M. A., Romero, D., and Wuest, T. 2018. A critical review of smart manufacturing & Industry 4.0 maturity models: Implications for small and medium-sized enterprises (SMEs). *Journal of manufacturing systems* 49: 194–214.

Müller, J., O. Buliga, and K. Voigt. 2018. "Fortune Favors the Prepared: How SMEs Approach Business Model Innovations in Industry 4.0." *Technological Forecasting and Social Change* 132: 2–17. doi:10.1016/j.techfore.2017.12.019.

Nannelli, M., F. Capone, and L. Lazzeretti. 2023. "Artificial Intelligence in Hospitality and Tourism. State of the art and Future Research Avenues." *European Planning Studies*, 1–20. doi:10.1080/09654313.2023.2180321.

Pagano, A., E. Carloni, S. Galvani, and R. Bocconcelli. 2021. "The Dissemination Mechanisms of Industry 4.0 Knowledge in Traditional Industrial Districts: Evidence from Italy." *Competitiveness Review* 31 (1): 27–53. doi:10.1108/CR-12-2019-0160.

Parcu, P., M. Rossi, N. Innocenti, and C. Carrozza. 2023. How Real Will the Metaverse Be? Exploring the Spatial Impact of Virtual Worlds. *European Planning Studies.* doi:10.1080/09654313.2023.2221323.

Porter, M., and J. Heppelmann. 2014. "How Smart, Connected Products are Transforming Competition." *Harvard Business Review* (Issue November).

Porter, M., and J. Heppelmann. 2015. "How Smart, Connected Products are Transforming Companies." *Harvard Business Review* (Issue October).

Scherrer, W. 2021. "Industry 4.0 as a 'Sudden Change': The Relevance of Long Waves of Economic Development for the Regional Level." *European Planning Studies* 29 (9): 1723–1737. doi:10.1080/09654313.2021.1963054.

Sultana, N., E. Turkina, and P. Cohendet. 2023. "The Mechanisms Underlying the Emergence of Innovation Ecosystems: The Case of the AI Ecosystem in Montreal." *European Planning Studies*, 1–23. doi:10.1080/09654313.2023.2185502.

Zukin, S. 2020. *The Innovation Complex: Cities, Tech, and the New Economy.* New York, NY: Oxford University Press.

The emergence of artificial intelligence in the regional sciences: a literature review

Luciana Lazzeretti, Niccolò Innocenti, Martina Nannelli and Stefania Oliva

Abstract

The article aims to analysing the literature on AI in regional sciences to understand the evolution of the topic in the field. Through a bibliometric analysis, it identifies the most publishing journals, most cited authors and relevant topics analysing more than 800 articles published between 1986 and 2020. Moreover, it reviews the most recent avenues of the literature analysing in-depth the content of 70 articles published in 2020 and 2021 in relevant journals in the field of innovation and regional science. From the analysis of the recent literature, six groups of topics emerge: Industry 4.0, smart cities, big data, AI and related technologies (robotization, IoT, augmented and virtual reality). The results confirm that AI is still an emerging topic in regional science and contribute to identifying the most intriguing issue and future research trends for developing this new research line.

1. Introduction

Over the last years, the entire society is living a paradigmatic change due to the run of innovations related to Artificial Intelligence (AI). We are in the summer of AI, and we are experiencing an unprecedented exponential growth of data both in terms of data production and exchange and in their speed and accessibility, largely sustained by the Internet, the use of social networks and digital platforms. We are going through a period of strong change characterised by what Pasquale (2015), referring to cybernetics and the concept of black box, defines the *black box society*. In the last few decades, several definitions of AI have been used: some broader and inclusive, others narrower and exclusive; some weaker and problem-solving oriented, others stronger for deep learning (Acemoglu and Restrepo 2018; Cristianini 2021). During the last years, we shifted from knowledge-driven to data-driven AI, while the emerging and rising data sciences are attracting many disciplines, not only scientific but also humanistic, economic and social. AI is considered by many scholars as a disruptive innovation, able to destroy and replace services and products, but also generate new ones (Girasa 2020). A paradigm-shifting innovation that has led to a real change in the techno-economic paradigm (Perez 2002) affecting not only the economy but also society itself.

The disruptiveness of AI is expanding exponentially and an increasing number of articles have been written to explore its paradigm, fields of application and impact on research areas (McAfee and Brynjolfsson 2012; Cristianini 2014; Amabile 2019; Floridi and Cowls 2019; Acemoglu and Restrepo 2020; Cristianini and Scantamburlo 2020; Lazzeretti 2020). In 2018, the economists Furman and Seamans (2019) refer to AI as a '*loose term [...] used to describe a range of advanced technologies that exhibit human-like intelligence, including machine learning, autonomous robotics and vehicles, computer vision, language processing, virtual agents and neural networks*'. However, we still know little concerning the impact of this new paradigm in the field of social sciences and, particularly, in regional sciences.

In a complex and sophisticated bibliometric analysis conducted using Microsoft Academic Graph, Frank et al. (2019) have analysed the evolution of AI research from 1950 to now in the different disciplinary fields and have highlighted precisely this gap. The authors found two main trends regarding the citations: an early period in which the main reference for data scientists was linked to the contributions of the humanities (philosophy, ethics, psychology, etc.), then a second period which was mainly driven by mathematicians, physicists and technologists. Meanwhile, other disciplines are few, particularly those related to social sciences. Some efforts have been recently done in management studies for summarizing the current literature on AI in specific domains, such as strategic management and AI (Keding 2021), innovation management (Haefner et al. 2021) or new emerging fields, such as AI and sustainability (Di Vaio et al. 2020). However, the relationship between social science and AI is clear, especially concerning the role that it can assume in decision processes at the economic and social levels and the impact that AI can have on the life of individuals and communities, which is increasingly pervasive from the institutional to the economic, political, cultural and social level.

Based on such considerations, the present article wants to enlarge the debate on the role of AI in the social sciences with particular attention to the regional sciences, the role of territories, innovations and economic development. Analysing the literature on AI in regional sciences, the article aims to understand the evolution of the topic in the field by identifying the journals where the debate is most vibrant, the main authors and research themes and debating the most recent and discussed topics.

The analysis of the literature follows two approaches. First, the article applies a bibliometric analysis for identifying most publishing journals, most cited authors and relevant topics in articles concerning regional science published between 1986 and 2020. Secondly, it reviews the most recent avenues of the literature analysing the content of articles published in 2020 and 2021 in the most important journals in the field of innovation and regional science. Results may contribute to identifying the relevant topics and future lines of studying the relationship between AI and regional science.

The article is structured as follows. Section 2 explains in detail the methodology adopted for the literature review. Section 3 presents the results of the bibliometric analysis, while Section 4 discusses the relevant topics that emerged in high-ranked journals in regional sciences and innovation. Finally, conclusions draw suggestions for future research that emerged from the analysis of the literature.

2. Methodology

Over the last few years, the literature review has raised its importance as a methodology for analysing several themes in regional sciences (Fröhlich and Hassink 2018; Zhu et al., 2019). This has resulted in the emergence of increasingly sophisticated and diversified tools for revising scientific literature. Bibliometric analyses have been proposed to successfully investigate topics of innovation and regional studies fields to understand the spread and distribution of scientific publications (Fagerberg and Verspagen 2009; Sedita, Caloffi, and Lazzeretti 2020). Other scholars have applied a systematic review to explore wide and blurry concepts, such as related variety (Content and Frenken 2016). Finally, some others have combined bibliometric analysis and content analysis for investigating the conceptual links between the circular economy and sharing economy (Henry et al. 2021).

The literature review applied in this article follows two approaches. First, a bibliometric analysis is conducted for identifying most publishing journals, most cited authors and relevant topics in articles concerning AI in regional sciences, published between 1986 and 2020. Secondly, the most recent avenues of the literature are revised by analysing the content of articles published in 2020 and 2021 in the most important journals in the field of innovation and regional sciences. Figure 1 summarises the approach followed for selecting the articles.

In the first step of the analysis, data has been collected from the ISI - Web of Science database, considering a period that goes from 1986 to 2020 through several steps. First, keywords have been selected thanks to consulting the work of expert scholars in the field and exploratory readings of the literature. We have included topics closely related to the concepts of AI such as machine learning, big data, autonomous vehicles, internet of things (IoT), robotics, and industry 4.0 (I4.0). Moreover, some specific criteria have been applied for selection: *source*, to have high-quality peer-reviewed pieces of research, only scientific articles have been selected; *language*, to correctly understand the contents of the articles, only articles published in the English language have been collected; *research domain*, considering the aim of the analysis, only articles that pertain to

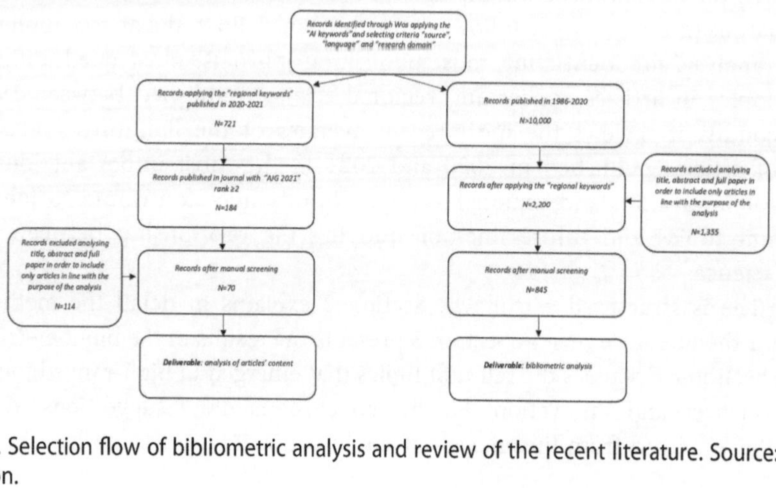

Figure 1 . Selection flow of bibliometric analysis and review of the recent literature. Source: Authors' elaboration.

economics, regional economics and economic geography disciplines (e.g. business, economics, geography, management, social issue and urban studies) have been selected.

This first research round has allowed us to collect more the 10,000 papers. Further research has provided for the introduction of those keywords that identify the regional science topics related to regional and urban studies (i.e. city, cluster, region and so on) selecting around 2,200 contributions.

Finally, accurate cleaning of the database has involved the reading of the abstracts (or the entire article when necessary) to exclude duplicates and those papers outside the focus of our research. The final database consisted of 845 articles that constitute the base of the bibliometric analysis.

In the second step of the analysis, to investigate the most recent debate in the literature, the same criteria and keywords have been applied to select articles in WoS published in 2020 and 2021. To restrict the research to relevant literature, selection has involved only articles published in the most relevant international scientific journals of the regional sciences based on the classification proposed by the Academic Journal Guide (AJG) 2021. The AJG is published every three years, assigns a score to journals based upon peer review, editorial and expert judgments following the evaluation of many hundreds of publications and is informed by statistical information relating to citation. To make a selection as inclusive as possible, we have selected the articles published in journals relating to the fields 'Regional studies, Planning and Environment' and 'Innovation'. Furthermore, the threshold chosen to select the most relevant journals was a minimum rank of 2 (on a ranking ranging from 1 to 4*).

Finally, we have refined our selection, through an in-depth reading of abstracts and the entire articles. The main selection criteria have included those articles where the city/regional dimension was significant, theoretically or empirically. The review was conducted on the content of 70 articles, for identifying the emerging themes in the discussion of AI in regional sciences and the most important line for future research in the very recent literature. For this purpose, for each article the following categories have been identified, when possible: 1) the typology of the article, distinguishing between theoretical and empirical; 2) the aim and research questions; 3) the research methodology; 4) geographical dimension or sector of application of the analysis; 5) the theoretical framework and research strand; 6) the definition of AI adopted by the authors; 7) main findings and suggestions for future research; 8) the main topic discussed in AI and regional sciences. The final step involved a clusterisation of the articles according to the main topic that emerged from the analysis.

3. The evolution of publications of AI and regional sciences

The preliminary results of the bibliometric analysis of AI in regional science offer an overview of the evolution of the topic over time. As a first result, we find a confirmation of the analysis of Frank et al. (2019). According to their analysis, indeed, the contribution of AI in the fields of economics and social sciences has been very modest over time, confirming a delay in the economic and social scientific debate compared to scientific and humanities fields. Looking at the results, the first appearance of AI concepts in the field is in 1986. Moreover, three different periods in the evolution of studies on AI in regional science can be identified in parallel with the growth of the number of publications (Figure 2) and citations (Figure 3). The first period coincided with the *birth* of the concept, before 2014 (less than 20 published articles every single year). The second *exploration* period goes from 2015 to 2017 in which a phase of growth in publication occurs (between 20 and 100 articles published). Finally, the *explosion* period after 2018 (more than 100 published articles every single year).

These results may depend on the nature of these kinds of innovations. They had first involved products and production processes and only subsequently had an impact on the social dimension, in terms of decision-making processes, both in organizations and in social contexts, partly as a consequence of the rapid improvement of product performance and increase in applications (Russell, Dewey, and Tegmark 2015; Wamba et al. 2021).

Another important result concerns the journals in which contributions have been published (Table 1). The main journals publishing on the topics are *Computers Environment and Urban Systems* which deals with innovative information technology (IT) research topics on urban systems and *Technological Forecasting and Social Change* which deals with topics of innovation, evaluation and technological forecasting to

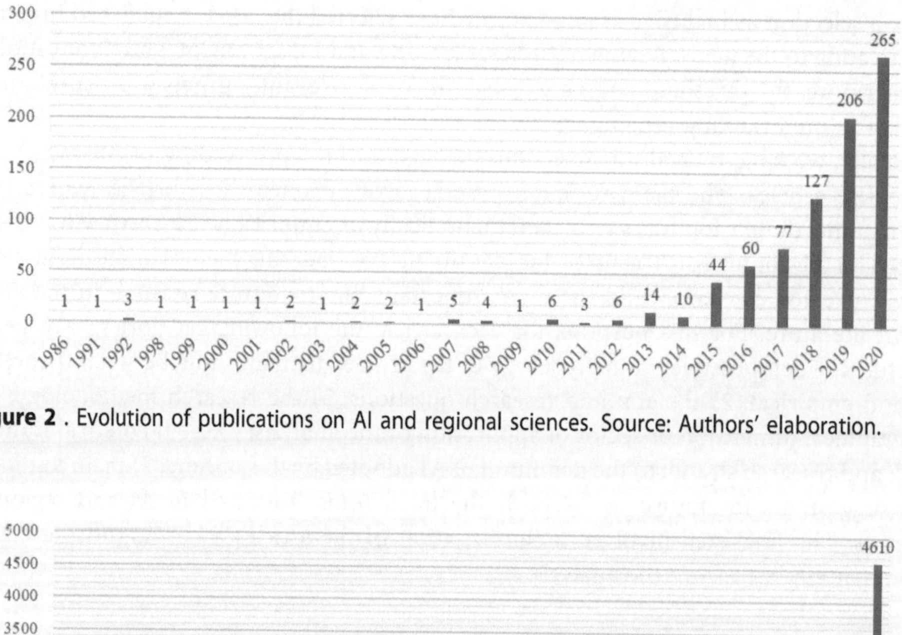

Figure 2 . Evolution of publications on AI and regional sciences. Source: Authors' elaboration.

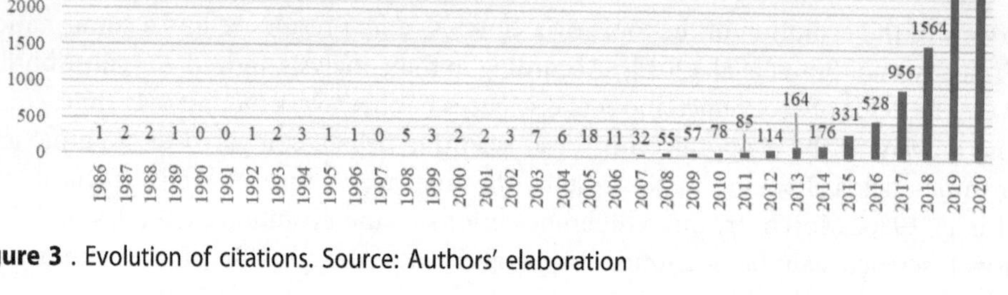

Figure 3 . Evolution of citations. Source: Authors' elaboration

Table 1 . The 30 most publishing journals.

N.	JOURNAL	N° of Articles
1	COMPUTERS ENVIRONMENT AND URBAN SYSTEMS	72
2	TECHNOLOGICAL FORECASTING AND SOCIAL CHANGE	51
3	CITIES	46
4	INTERNATIONAL JOURNAL OF GEOGRAPHICAL INFORMATION SCIENCE	39
5	ENVIRONMENT AND PLANNING B-URBAN ANALYTICS AND CITY SCIENCE	23
6	TRANSPORTATION RESEARCH PART A-POLICY AND PRACTICE	21
7	TRANSACTIONS IN GIS	20
8	URBAN SCIENCE	19
9	BIG DATA & SOCIETY	19
10	ANNALS OF THE AMERICAN ASSOCIATION OF GEOGRAPHERS	13
11	JOURNAL OF TRANSPORT GEOGRAPHY	13
12	LANDSCAPE AND URBAN PLANNING	11
13	TRANSPORT POLICY	11
14	ENVIRONMENT AND PLANNING A-ECONOMY AND SPACE	11
15	HABITAT INTERNATIONAL	11
16	URBAN STUDIES	11
17	URBAN FORESTRY & URBAN GREENING	10
18	RESEARCH IN TRANSPORTATION ECONOMICS	9
19	TOURISM MANAGEMENT	9
20	APPLIED GEOGRAPHY	8
21	CAMBRIDGE JOURNAL OF REGIONS ECONOMY AND SOCIETY	7
22	SOCIAL INDICATORS RESEARCH	7
23	PROGRESS IN HUMAN GEOGRAPHY	7
24	REGIONAL SCIENCE AND URBAN ECONOMICS	6
25	URBAN PLANNING	6
26	URBAN GEOGRAPHY	6
27	JOURNAL OF DESTINATION MARKETING & MANAGEMENT	6
28	JOURNAL OF URBAN TECHNOLOGY	6
29	INTERDISCIPLINARY DESCRIPTION OF COMPLEX SYSTEMS	6
30	JOURNAL OF URBAN PLANNING AND DEVELOPMENT	6

Source: Authors' elaboration

underline how innovation and ITs are the topics that mainly deal with these issues. To be noted is a growing interest of economic scholars, in particular, in the studies of management, geography and urban studies. Given the aim of our research and the selection criteria, many journals fall in the disciplines of regional sciences (for example, *Cities*). It should be noted that there are also two journals in the field of management and tourism, *Tourism Management* and *Journal of Destination Marketing & Management*, showing that the topic finds fertile ground in these disciplines.

Table 2 shows that many contributions are from authors affiliated with institutions based in the US and China, countries where this new discipline has firstly developed. In addition, these countries represent also examples of the faster development of AI technologies in several productive domains. However, the presence of authors affiliated with European institutions, especially from Italy, Belgium and Spain has recently grown and is reaching relevant positions in terms of average citations, which can be considered indicative of the influence of their research.

Table 3 shows the most cited articles on AI in regional sciences. In this regard, it is important to note that among the most cited, many addresses the theme of the smart city and also big data, underlining how these two are focal topics in the literature. This is also confirmed by the results of Figure 4, which shows the main topics emerging from the articles' keywords. Big data is the most discussed topic and with others, as machine learning, deep learning, data mining, artificial neural networks and text

Table 2 . The 10 most cited authors and University affiliation.

N.	AUTHORS NAME	AV. CITATIONS PER YEAR	COUNTRY OF AFFILIATION
1	Zook, Matthew	99,91	USA
2	Shelton, Taylor	80,54	USA
3	Yao, Yao	72,14	China
4	Del Giudice, Manlio	57,42	Italy
5	Li, Xia	54,86	China
6	Zhang, Jinbao	51,93	China
7	Poorthuis, Ate	49,41	Belgium
8	Mai, Ke	48,73	China
9	Gutierrez, Javier	47,31	Spain
10	Liang, Zhaotang	46,73	China

Source: Authors' elaboration

Table 3 . The 10 most cited articles.

N.	AUTHOR	ARTICLE TITLE	JOURNAL	AVERAGE CITATIONS PER YEAR	PUBLICATION YEAR
1	El-Kassar, AN; Singh, SK	*Green innovation and organizational performance: The influence of big data and the moderating role of management commitment and HR practices*	TECHNOLOGICAL FORECASTING AND SOCIAL CHANGE	45,33	2019
2	Shelton, T; Zook, M; Wiig, A	*The 'actually existing smart city'*	CAMBRIDGE JOURNAL OF REGIONS ECONOMY AND SOCIETY	35,14	2015
3	Ma, XL; Liu, CC; Wen, HM; Wang, YP; Wu, YJ	*Understanding commuting patterns using transit smart card data*	JOURNAL OF TRANSPORT GEOGRAPHY	29	2017
4	Bresciani, S; Ferraris, A; Del Giudice, M	*The management of organizational ambidexterity through alliances in a new context of analysis: Internet of Things (IoT) smart city projects*	TECHNOLOGICAL FORECASTING AND SOCIAL CHANGE	28,75	2018
5	Chen, TD; Kockelman, KM; Hanna, JP	*Operations of a shared, autonomous, electric vehicle fleet: Implications of vehicle & charging infrastructure decisions*	TRANSPORTATION RESEARCH PART A-POLICY AND PRACTICE	27,5	2016
6	Sadowski, J	*When data is capital: Datafication, accumulation, and extraction*	BIG DATA & SOCIETY	27,33	2019
7	Allamm Z; Dhunny, ZA	*On big data, artificial intelligence and smart cities*	CITIES	27	2019
8	Netzer, O; Feldman, R; Goldenberg, J; Fresko, M	*Mine Your Own Business: Market-Structure Surveillance Through Text Mining*	MARKETING SCIENCE	25,8	2012
9	Yao, Y; Li, X; Liu, PH; Liang, ZT; Zhang, JB; Mai, K	*Sensing spatial distribution of urban land use by integrating points-of-interest and Google Word2Vec model*	INTERNATIONAL JOURNAL OF GEOGRAPHICAL INFORMATION SCIENCE	25,8	2017
10	Shelton, T; Poorthuis, A; Zook, M	*Social media and the city: Rethinking urban socio-spatial inequality using user-generated geographic information*	LANDSCAPE AND URBAN PLANNING	24	2015

Source: Authors' elaboration

Figure 4 . Relevant Topics in regional sciences. Source: Authors' elaboration

mining. These are methodologies used for the analysis of big data. In addition, another related topic is social media (e.g. Twitter) which is both an online environment where people interact, and a source of data. Other relevant topics are the smart city, urban planning and design and land use. The smart city is connected to other topics as automated urban mobility (autonomous vehicles, human mobility), smart infrastructure and governance. Finally, also I4.0 is an emergent theme connected to AI as well as smart manufacturing, smart automation, work automation and the general topic of digitalization of industries.

4. The future avenues of AI in regional sciences

A further step in this survey is represented by the analysis of the recent literature. The review is based on the articles published in the years 2020 and 2021 in journals with a rank of 2 or higher according to the AJG 2021 list and among the 'Regional studies, Planning and Environment' and 'Innovation' disciplines. The final selection counts 70 articles. Through an in-depth reading of the articles, we divided them into 5 groups based on the main topic developed. For each topic, the review has focused on how the literature defines the concept of AI, the main theoretical framework, results and challenges for AI in regional sciences.

4.1 Industry 4.0

A consistent share of the contributions (25 articles) pertains to the topic of Industry 4.0 (I4.0). The scientific journals where the debate is liveliest on the topic among the contributions identified are: *Regional Studies* (10 articles), followed by *Technological Forecasting and Social Change* (7 articles) and *European Planning Studies* (6 articles). Almost the totality of the articles affords the topic of I4.0 through empirical analysis (21 articles), while a residual part develops it theoretically (2 papers) or both, exploring theoretical issues and describing also empirical cases (2 papers). Among the methodologies applied for analysing the topic, these range from case studies to content analysis and statistical analysis, while concerning the geographical dimension of the analysis, it is possible to underline a tendency to conduct analyses at the European level (15 papers) rather than to a more fine-grained level, both in terms of European regions and single countries.

The group of articles develops the literature on I4.0 within different theoretical strands and addresses several topics. While part of the identified contributions frames their analyses in the Economic Geography literature (Balland and Boschma 2021; De Propris and Bailey 2021; Fromhold-Eisebith et al. 2021; Labory and Bianchi 2021), others develop knowledge in the regional innovation policy context (Lepore and Spigarelli 2020; Baker, Gaspard, and Zhu 2021; Sandulli, Gimenez-Fernandez, and Rodriguez Ferradas 2021) and a part focuses on the topic of value creation (Busch et al. 2021; Jeannerat and Theurillat 2021). While a residual part of the studies is more focused on the business dimension, according to different arguments such as product innovation (Kahle et al. 2020) or corporate social responsibility (Siltori et al. 2021).

It is relevant to mention that a large part of the articles (Bilgen 2021; Capello and Lenzi 2021; Corradini, Santini, and Vecciolini 2021; Siltori et al. 2021) trace the birth of the concept of I4.0 to the German industrial policy, in particular with the Hannover Fair in 2011 (Kagermann, Wahlster, and Helbig 2013). Some of the identified contributions refer to I4.0 generically in terms of a digitisation path of manufacturing (Hervás-Oliver 2021) that is changing the production model in firms and along the value chain (Barzotto and De Propris 2021) and considering that digitisation, smart industry, smart factory and smart manufacturing are all related concepts.

However, there is no unanimous consensus on specific technologies that pertain to the range of I4.0. This result in a fragmented contribution following a broader or narrower approach to the technologies. De Propris and Bailey (2021) focus their attention on the Fourth industrial revolution (FIR) where a new technological paradigm emerges among three different domains. The first is green technologies; the second involves digital technologies such as ICT and mobile technologies, additive manufacturing or 3D printing, artificial intelligence (AI), cloud computing, big data analytics, IoT, advanced robotics, sensoring, space technology and drones; the third relates to new materials (Culot et al. 2020). In analysing what capabilities drive the diversification in the I4.0 domains, Balland and Boschma (2021) conduct an analysis of the literature and identify 10 I4.0 technologies: (1) additive manufacturing; (2) AI; (3) augmented reality; (4) autonomous robots; (5) autonomous vehicles; (6) cloud computing; (7) cybersecurity; (8) machine tools; (9) quantum computers; (10) system integration. While Duman and Akdemir (2021) identify the technological components of I4.0 in cyber physical systems, IoT, big data, cloud computing, 3D printer/print, robotic applications, and augmented reality. Some overlaps exist among the categories and is possible to identify a core. However, the definition of borders remains under discussion and we can expect further developments also according to the technological advancements. Despite these efforts in clarifying the blurry definitional aspects of the I4.0 concept, the identified articles evidence that the literature on the topic is still in its infancy and requires further development. This is especially related to the drivers that may favour (or hinder) the transition toward I4.0 also by comparing the results of the existing analyses with further investigation in other regions. Moreover, the existing literature gives scarce attention to the topic of I4.0 in developing countries, while the main empirical applications are limited to developed countries. This has probably to do with the level of development required to act a transition to the I4.0 and to the relevant policy and economic efforts needed to invest in these new and complex technologies.

The connection with the contribution of AI to regional science emerges in terms of the importance that related technologies are gaining for the transformation of industries, business and organization models and work and the degree through which this transformation is related to locally embedded resources in a path-dependent process.

4.2 Smart cities

Among the identified contributions, it emerges that several articles (24 articles) discuss the smart city topic, mostly in the journals *Technological Forecasting and Social Change* (11 articles) and *Cities* (8 articles). The largest part of the articles gives an empirical contribution (13 articles) while a few less a theoretical one (10 articles) or both (1 article).

From the in-depth reading of the 24 recent articles, the concept of smart city is still controversial and has not yet found a univocal definition. Despite the differences in terms of definitions, three common dimensions can be identified: technological, human and institutional (Ranchod 2020).

First of all, the smart city concept is defined around the use of smart services using Information and Communication Technologies (ICTs) to enhance smart infrastructure and city performance through more effective and efficient infrastructure operations and management (Rice and Martin 2020; Chang 2021; Lim, Cho, and Kim 2021). Adler and Florida (2021) extensively discussing the economic geography of urban technology (urban tech and start-up enterprises) suggest that the geography of urban tech is shaped by the innovative capabilities of urban areas and they define the smart city as the environment where human capability social capital and informational technology meet (p.1788). The concept is supported by the IoT, the use of big data, social media analytics, open innovation and crowdsourcing (Gupta, Panagiotopoulos, and Bowen 2020), cloud computing, remote sensing, geographic information systems, and global positioning system technologies (Yan, Liu, and Tseng 2020).

Secondly, technological development is needed to improve the effectiveness of many services that directly affect the quality of citizen life such as transportation, health care, communication and energy supplies and offer better methods to address some urban challenges such as urban sustainability (Kummitha 2020; Rice and Martin 2020; Staletić et al. 2020; Liu et al. 2020). An ever-increasing relevance in the development and management of the smart city concerns public and private transport which is one of the most relevant topics. Larson and Weihua (2020) adapt the classic monocentric city model to discuss the effects of the autonomous vehicle on sprawl, energy consumption, and housing affordability. While many authors have addressed the issue generically (Araral 2020; Goel, Yadav, and Vishnoi 2021), some have focused their research on specific countries such as China and New York (Mohamed et al. 2020; Yan, Liu, and Tseng 2020; Iqbal et al. 2020).

The third set includes smart governance to enable a more efficient and integrated urban management (Caprotti and Liu 2020; Kummitha 2020; Kandt and Batty 2021). Therefore, the smart city concept, with the synergistic support of new technologies, can be considered a smart connected environment for better efficiency and management on multiple levels (Araral 2020; Chang 2021; Shamsuzzoha et al. 2021;).

The fuzziness of the concept is coupled with a variety of methodologies used to investigate it. Different qualitative and quantitative methodologies have been applied. Some scholars have explored the topic through qualitative literature review trying to understand the dynamics and factors composing the smart city (Caprotti and Liu 2020; Chang 2021; Shamsuzzoha et al. 2021), while others have applied quantitative statistical methodologies (Julsrud and Krogstad 2020; Larson and Weihua 2020; Araral 2020), big data analysis (Monachesi 2020) and others advanced methodologies such as text mining and social network analysis (Monachesi 2020). Other contributions followed a case study research approach, using both qualitative and quantitative methods, allowing for analysing of specific contexts such as Oslo and Tallin (Julsrud and Krogstad 2020), New York (Iqbal et al. 2020) and the Netherlands (Monachesi 2020) which are implementing the structure of the smart city. But also of developing countries and this is a clear difference compared to the studies of the first group (I4.0), meaning cities such as Nairobi (Chambers and Evans 2020) and secondary cities of South Africa (Ranchod 2020), to understand the paradigm's development potential.

Concerning regional development, AI tools are necessary to transversally build a whole smart environment, connect all the elements and enable the complete use of many technologies involved in the implementation of a smart city environment. Moreover, some ethical issues emerge in particular concerning centralized control, data privacy, biased data and inequalities among people whose trade-off between benefits and costs must be evaluated (Caprotti and Liu 2020; Mohamed et al. 2020; Chang 2021).

Finally, the avenues for future research underlines the priority of investigating further the phenomenon (Shamsuzzoha et al. 2021) both through the use of new methodologies and by expanding the analysis to different realities (Iqbal et al. 2020; Kumar et al. 2020), to identify new effective governance and also management possibilities for cities (Ranchod 2020; Gupta, Panagiotopoulos, and Bowen 2020) and to build a new research agenda to challenge smart city technological and ideological developments (Chambers and Evans 2020; Kummitha 2020).

4.3 Big data and social media

Searching for the links between AI and regional science, the topic of big data emerges as one of the most relevant (11 articles). Among these articles, 9 contributions are published in *Cities* underlining the connection with the topics of urban development and land use, the others are published respectively in *Regional Studies*, *Urban Studies* and *Regional Sciences and Urban Economics*. The articles have a mostly empirical nature (11 articles), excluding one that discusses the topic theoretically. The methodologies applied use mostly quantitative-statistical and econometric methods such as the Tencent location big data for network analysis and spatial analysis (Hui et al. 2020), machine learning and computer vision algorithms, a weighted stochastic block model (WSBM) (Wang and Vermeulen 2021), toponym co-occurrence and improved field strength model (Wu et al. 2020), and the random forest method and dynamic topic modelling approach (Kim et al. 2021).

The predominance of empirical studies is due in the nature of big data, a collection of information data so extensive in terms of volume, speed and variety that specific analytical technologies and methods are required for the extraction of their value or knowledge. Indeed, the definition of big data is linked to their ability to describe and/or estimate social, economic and political phenomena (Yang, Wu, and Wang 2021). In particular, it is connected to their power to estimate a large amount of data from multiple sources, for a variety of reasons. For example, big data are widely used to obtain data on citizens and consumers for profiling purpose for both private (advertising) and public (virus tracing) interests (Indaco 2020); to estimate the spatial and geographical evolution of cities, urban agglomeration (Wu et al. 2020), land use (He et al. 2020; Liu et al. 2020) and urban traffic patterns (Hui et al. 2020).

As already presented, the use of big data needs a large source of data in terms of volume; for this reason, this topic is frequently overlapped with social media, one of the most frequent sources of huge volumes of data (Fang et al. 2020; Liu et al. 2020; Wang and Vermeulen 2021; Olson et al. 2021). These data provide a basin of information in real-time capable of exploring the geography of time–space aggregate human activity (Liu et al. 2020), analysing the city network for urban agglomeration (Fang et al. 2020) or the extent to which the built environment impacts the survival rate of neighbourhood-based social organizations (Wang and Vermeulen 2021) and examining the relations among neighbourhoods using social media review (Olson et al. 2021). Moreover, the theoretical framework of the big data and social media finds ground in urban studies discipline.

Big data and the tools for exploring it offer further research opportunities and scholars agree on the need to improve studies and techniques on them. They underline the need to refine the research and methodologies because some relevant doubts or obstacles emerge from the use of such data, mainly concerning methods that are not yet sufficiently refined (Yang, Wu, and Wang 2021; Kim et al. 2021) and the need to develop modern computational tools moving from description to prediction (Olson et al. 2021).

4.4 Artificial intelligence

A group of articles deals with the topic of AI using a more general perspective. These are 4 articles, of which 2 use an empirical approach and 2 a theoretical one. The articles are published in *Cambridge Journal of Regions, Economy and Society* (2 articles), *Regional Studies* (1 article), and *European Planning Studies* (1 article).

For a theoretical perspective, 2 articles pertain to the literature on Evolutionary Economic Geography (EEG) and want to explain how the industry of AI emerges, its conditions and its characteristics. The first one (Buarque et al. 2020) uses patent data to explore the interaction of AI and geography across European regions through the methodology of knowledge space. It shows that the regions developing innovations related to the information technology sector are most likely to develop also innovations in the AI field. Furthermore, the most innovative regions in AI are those for which this domain is most integrated into the knowledge space. Doloreux and Turkina (2021) carry out a case study analysis for exploring the conditions through which the area of Montreal developed its AI industry following the path creation approach. The article recognizes the need for an interrelation between several factors in fostering the development of the AI industry in a geographic area. First of all, the existence of a network of companies in ICT, as well as the promoting role of entrepreneurs and non-economic organizations that support the sector, both through global relations and through multi-scale political initiatives.

The other contributions discuss the open questions raised by the advent of AI, recalling the definition of AI by the seminal work of Russell and Norvig (2003). The first one, Dignam (2020), identifies some of the challenges that arise from the adoption of AI in the context of its regulation. The first concern is related to the excessive cost of producing a high-quality AI model. Moreover, problems can arise from using data that are not representative of the entire society, generating bias, at the expense of underrepresented categories of the society. A further problem arises from the inability to fully understand how certain types of AI work and therefore the inability to explain certain results generated by resorting to AI. Furthermore, AI processes give outcomes that are a function of human design, this could favour the private rather than the public interest. Such questions, among others, require urgently a regulatory framework. After discussing the birth of AI and how it has transformed society, with particular attention to culture and creativity and culture-driven development models, Lazzeretti (2020) discusses the possible risks faced by the new 'algorithmic society'. Even in this case, the possibility of bias generated by decision-making processes based on AI is underlined. Alongside these, emphasis is placed on the risks of memory loss and fundamental rights, the alienation of the individual and the impoverishment of natural language and the theme of inequalities - economic, digital and recognition - is discussed.

This group of articles, on the one side, try to explicate the conditions for the birth of the AI industry and the competitive advantages for its flourishing development at the regional level. On the other side, they highlight the efforts that social and regional scientists should done for studying AI as a social phenomenon, underlining several unsolved issues and risks connected to the adoption of AI in many fields of human life.

4.5 Robotization, ioT, augmented and virtual reality

The last group of articles includes contributions that investigate specific technologies of AI that have been investigated residually in the literature of AI and regional science. The group includes 5 articles published in *Technological Forecasting and Social Change*, *Technovation*, *Cambridge Journal of Regions Economy and Society*, *Cities* and *Urban studies* (1 article each). Among the identified contributions, two address questions related to the topic of robotization, two deal with the use of augmented and virtual reality technologies and one concerns the analysis of IoT technologies.

Concerning the topic of robotization, the articles offer two interesting perspectives on the application of robots in the industrial dimension and city context. Gentili et al. (2020) empirically investigate the effects of the use of robotics technology on productivity, employment and wages analysing the case of Denmark, France, Germany, Italy, the UK and the US. The results of the analysis show that this relationship can be positive or negative in different economic structures, in particular in terms of economic specializations. The process of labour robotization is time-sensitive and related to the stage of development of a country and its socio-institutional structure. The article by While, Marvin, and Kovacic (2021) deals with the topic of robotization in the literature of urban studies, exploring the adoption of autonomous social robots in cities. In particular, the article offers a framework for analysing the application of robotic systems to specific dimensions of urban life and describes three cases of robotic experimentation in San Francisco, Tokyo and Dubai. Despite the adoption of robots in the cities' context is still at an early stage, the cases evidence substantial differences. If cities want to start a process of 'infrastructuralisation' in terms of robotically augmented cities, different dimensions of urban planning, regulation and design should co-evolve as well as the human-robotic interaction.

Among the identified contributions, two papers empirically afford the use of virtual and augmented reality namely in urban infrastructure management and tourism. In the context of urban planning, especially in the management of infrastructures, Ma et al. (2020) develop a tool based on virtual reality for communicating the complexity of interrelated spatial systems data. In their analysis, they show that virtual reality may represent an innovation by cross-fertilization developed in the gaming industry and applied to spatial planning through three-dimensional scenarios from overlapping spatial datasets, such as 3D GIS. This can promote a better understanding of the urban environment for the stakeholders involved in the decision process of cities planning. The article of Huang and Liu (2021), instead, applies the use of augmented reality to the case of the tourism industry in order to increase the love branding of a tourism destination. The analysis shows that disruptive technologies, such as augmented reality, can help to humanize the digital experience and increase the brand love of tourism destinations. Moreover, they can replace the real experience when this cannot be performed by visitors, as in the case of restrictions due to Covid-19.

Finally, Bessagnet, Crespo, and Vicente (2021) discuss the specific case of IoT technologies through a historical event analysis applied to Toulouse, France. The authors frame their analysis in the context of Entrepreneurial Ecosystems to describe the evolution and scale of the technological trajectory in the area, identifying the initial conditions, main features of the regional environment and the dynamics of the technology and platform worldwide. The analysis highlights that some factors can explain the success of the technological path within the region and are related to the presence of a blockbuster, related activities, entrepreneurial capacity and start-up activities and the support by the local government in terms of incentives and infrastructures.

All these articles showed clear evidence of the relevant relationships with the themes of regional sciences, both in terms of the possible robotization of the cities and in relation to the augmented experience tourists could have using augmented realities tools. In this respect, it should be noted that the full realization of these two processes cannot take place without the deployment of enabling technologies (such as 5G and AI) and the evolution of the IoT industry.

5. Conclusions

This article represents the first attempt to systematise the wide and complex literature on AI in social sciences focusing on a still underdeveloped but growing field: the regional sciences. The analysis is divided into two parts. It first applied a bibliometric analysis on AI and regional sciences in the period ranging from 1986 to 2020. This analysis shows how this field of research is booming particularly in the last three years involving an increasing number of journals and countries. Secondly, it developed a review on relevant topics in the most recent publications (2020-2021), grouping the contributions into five groups in a broad sense - I4.0, smart cities, big data, AI - and a final residual group that includes contributions that investigate specific technologies (robotization, IoT, augmented and virtual reality).

Besides the strong growth of AI themes and the even stronger acceleration in the last few years as well as their increasing relevance in the fields of social sciences and particularly regional sciences, what emerged clearly from the in-depth reading of these contributions and the systematization of the literature is the strict relations among the five topics identified. As already suggested regarding the IoT topic, all these fields of research, but most importantly their real applications, are strictly related to each other, and this is particularly evident when we approach this topic using a geographical perspective. When discussing the technologies *per se* the relation among them is much more latent. However, when we think of the real use and deployment of these processes or technologies the relation strongly appears in our arena. There could not be a full realization of the I4.0 without the development of the 'Internet of Things' (IoT) and the same applies to the smart city. At the same time, for autonomous vehicles, the use of big data and machine learning comes into play. This also emerges from empirical evidence, highlighting the role that a knowledge base in ICTs plays in enabling regions to diversify into AI technologies (Xiao and Boschma 2021). For this reason, the studies of specific themes related to AI and regional science require better firstly to define the AI concept as a whole, according to a regional science perspective. This clearly emerges from the analysis of the literature, recognizing uncertainty in what can be classified under the umbrella of AI, in terms of technologies but also concerning its theoretical definition.

Consistent with Frank et al. (2019), the proposed analysis confirms that AI is an emerging topic in the field of regional sciences. Also, recent literature suggests a predominance of the technical aspects of AI at the expense of research on the social dimension (Wamba et al. 2021). This could be explained by the speed by which this type of innovation was born and continues to develop. Indeed, it first involved products and production processes and subsequently had an impact on society, as in the case of decision-making processes or the development of cities. In the regional sciences, moreover, the most productive areas are I4.0 and smart cities. This may also depend on the thrust of industrial and urban policies on these themes. This strongly emerges from the analysis of the contributions on I4.0 which recall the 2011 Hannover Fair as the birthplace of the concept.

However, if policies have accelerated the development of studies in certain areas, many issues remain unsolved, fostering the development of future lines of research in the field of AI and regional sciences. As recent literature suggests (Cooke 2021), AI and related tools raise important questions concerning the privacy dimension of their applications, also at the territorial level. This is the case, for instance, of big data for smart cities, where the collection and production of information, if not carefully regulated, can lead to undemocratic behaviour and generate wide profit for private companies (Cooke 2020).

Moreover, despite the application of AI at several levels of industrial production, its whole impact on the economic development of places remains under-research (Russell, Dewey, and Tegmark 2015). A higher effort is required to investigate possible distortions caused by the adoption of such technologies in the labour and economic markets, in terms of transparency and asymmetric information.

Finally, in line with the analysed articles (Lazzeretti 2020; Balland and Boschma 2021), more research is required to investigate the consequences produced by the adoption of AI technologies for spatial inequalities. It seems important to study the role of giant companies in economic development, to assess the costs and benefits for the development of places. This avenue can be promising in giving insights to policymakers to understand how to answer to the change driven by AI.

This article sheds light on the still poorly investigated field of AI and regional science. However, it also has certain limitations. On the one side, it represents an exploratory investigation in the field and requires future in-depth studies. On the other side, despite the use of well-established bibliometric tools, bibliometric analysis is mostly descriptive. This is partially solved by the review that explores the content of recent articles in detail but which, on the contrary, it is based on a limited number of papers subjectively selected by authors.

Despite the acknowledged limitations, considering the novelty of the topic and the paucity of literature reviews investigating the relationship between AI and regional sciences, the analysis can provide valuable insights and suggest evidence-based research directions, both for academia and policymakers, for developing such an under-debated but intriguing field.

Disclosure statement

No potential conflict of interest was reported by the author(s).

ORCID

Luciana Lazzeretti http://orcid.org/0000-0002-9759-2289
Niccolò Innocenti http://orcid.org/0000-0001-8421-5479
Martina Nannelli http://orcid.org/0000-0003-2342-0051
Stefania Oliva http://orcid.org/0000-0003-2933-4795

References

Acemoglu, D., and P. Restrepo. 2018. "The Race Between Man and Machine: Implications of Technology for Growth, Factor Shares, and Employment." *American Economic Review* 108 (6): 1488–1542. doi:10.1257/aer.20160696

Acemoglu, D., and P. Restrepo. 2020. "The Wrong Kind of AI? Artificial Intelligence and the Future of Labour Demand." *Cambridge Journal of Regions, Economy and Society* 13 (1): 25–35. doi:10.1093/cjres/rsz022

Adler, P., and R. Florida. 2021. "The Rise of Urban Tech: How Innovations for Cities Come from Cities." *Regional Studies* 55 (10-11): 1787–1800. doi:10.1080/00343404.2021.1962520

Amabile, T. 2019. "Creativity, Artificial Intelligence, and a World of Surprises." *Academy of Management Discoveries* 6 (3).

Araral, E. 2020. "Why do Cities Adopt Smart Technologies? Contingency Theory and Evidence from the United States." *Cities* 106: 102873. doi:10.1016/j.cities.2020.102873

Baker, P. M., H. Gaspard, and J. A. Zhu. 2021. "Industry 4.0/Digitalization and Networks of Innovation in the North American Regional Context." *European Planning Studies* 29 (9): 1708–1722. doi:10.1080/09654313.2021.1963053

Balland, P. A., and R. Boschma. 2021. "Mapping the Potentials of Regions in Europe to Contribute to new Knowledge Production in Industry 4.0 Technologies." *Regional Studies* 55 (10-11): 1652–1666. doi:10.1080/00343404.2021.1900557

Barzotto, M., and L. De Propris. 2021. "The Value of Firm Linkages in the age of Industry 4.0: A Qualitative Comparative Analysis." *The Annals of Regional Science* 67 (2): 245–272. doi:10.1007/s00168-021-01047-0

Bessagnet, A., J. Crespo, and J. Vicente. 2021. "Unraveling the Multi-Scalar and Evolutionary Forces of Entrepreneurial Ecosystems: A Historical Event Analysis Applied to IoT Valley." *Technovation* 108: 102329. doi:10.1016/j.technovation.2021.102329

Bilgen, H. 2021. "A Global Comparison Methodology to Determine Critical Requirements for Achieving Industry 4.0." *Technological Forecasting and Social Change* 172: 121036. doi:10.1016/j.techfore.2021.121036

Buarque, B. S., R. B. Davies, R. M. Hynes, and D. F. Kogler. 2020. "OK Computer: The Creation and Integration of AI in Europe." *Cambridge Journal of Regions, Economy and Society* 13 (1): 175–192. doi:10.1093/cjres/rsz023

Busch, H. C., C. Mühl, M. Fuchs, and M. Fromhold-Eisebith. 2021. "Digital Urban Production: How Does Industry 4.0 Reconfigure Productive Value Creation in Urban Contexts?" *Regional Studies* 55 (10-11): 1801–1815. doi:10.1080/00343404.2021.1957460

Capello, R., and C. Lenzi. 2021. "Industry 4.0 and Servitisation: Regional Patterns of 4.0 Technological Transformations in Europe." *Technological Forecasting and Social Change* 173: 121164. doi:10.1016/j.techfore.2021.121164

Caprotti, F., and D. Liu. 2020. Emerging platform urbanism in China: Reconfigurations of data, citizenship and materialities.

Chambers, J., and J. Evans. 2020. "Informal Urbanism and the Internet of Things: Reliability, Trust and the Reconfiguration of Infrastructure." *Urban Studies* 57 (14): 2918–2935. doi:10.1177/0042098019890798

Chang, V. 2021. "An Ethical Framework for big Data and Smart Cities." *Technological Forecasting and Social Change* 165: 120559. doi:10.1016/j.techfore.2020.120559

Content, J., and K. Frenken. 2016. "Related Variety and Economic Development: A Literature Review." *European Planning Studies* 24 (12): 2097–2112. doi:10.1080/09654313.2016.1246517

Cooke, P. 2020. "Silicon Valley Imperialists Create new Model Villages as Smart Cities in Their own Image." *Journal of Open Innovation: Technology, Market, and Complexity* 6 (2): 24. doi:10.3390/joitmc6020024

Cooke, P. 2021. "Three Disruptive Models of new Spatial Planning: "attention", "Surveillance" or "Sustainable" Capitalisms?" *Journal of Open Innovation: Technology, Market, and Complexity* 7 (1): 46. doi:10.3390/joitmc7010046

Corradini, C., E. Santini, and C. Vecciolini. 2021. "The Geography of Industry 4.0 Technologies Across European Regions." *Regional Studies* 55 (10-11): 1667–1680. doi:10.1080/00343404.2021.1884216

Cristianini, N. 2014. "On the Current Paradigm in Artificial Intelligence." *AI Communications* 27 (1): 37–43. doi:10.3233/AIC-130582

Cristianini, N. 2021. "Shortcuts to Artificial Intelligence." In *Machines We Trust*, edited by M. Pelillo, and T. Scantamburlo, 11–26. Cambridge, MA: MIT Press.

Cristianini, N., and T. Scantamburlo. 2020. "On Social Machines for Algorithmic Regulation." *AI & Society* 35: 645–662. doi:10.1007/s00146-019-00917-8

Culot, G., G. Nassimbeni, G. Orzes, and M. Sartor. 2020. "Behind the Definition of Industry 4.0: Analysis and Open Questions." *International Journal of Production Economics* 226: 107617. doi:10.1016/j.ijpe.2020.107617

De Propris, L., and D. Bailey. 2021. "Pathways of Regional Transformation and Industry 4.0." *Regional Studies* 55 (10-11): 1617–1629. doi:10.1080/00343404.2021.1960962

Di Vaio, A., R. Palladino, R. Hassan, and O. Escobar. 2020. "Artificial Intelligence and Business Models in the Sustainable Development Goals Perspective: A Systematic Literature Review." *Journal of Business Research* 121: 283–314. doi:10.1016/j.jbusres.2020.08.019

Dignam, A. 2020. "Artificial Intelligence, Tech Corporate Governance and the Public Interest Regulatory Response." *Cambridge Journal of Regions, Economy and Society* 13 (1): 37–54. doi:10.1093/cjres/rsaa002

Doloreux, D., and E. Turkina. 2021. "New Path Creation in the Artificial Intelligence Industry: Regional Preconditions, new Actors and Their Collective Actions, and Policies." *Regional Studies* 55 (10-11): 1751–1763. doi:10.1080/00343404.2021.1919610

Duman, M. C., and B. Akdemir. 2021. "A Study to Determine the Effects of Industry 4.0 Technology Components on Organizational Performance." *Technological Forecasting and Social Change* 167: 120615. doi:10.1016/j.techfore.2021.120615

Fagerberg, J., and B. Verspagen. 2009. "Innovation Studies—The Emerging Structure of a new Scientific Field." *Research Policy* 38: 218–233. doi:10.1016/j.respol.2008.12.006

Fang, C., X. Yu, X. Zhang, J. Fang, and H. Liu. 2020. "Big Data Analysis on the Spatial Networks of Urban Agglomeration." *Cities* 102: 102735. doi:10.1016/j.cities.2020.102735

Floridi, L., and J. Cowls. 2019. "A Unified Framework of Five Principles for AI in Society." *Harvard Data Science Review* 1 (1): 1–15.

Frank, M. R., D. Wang, M. Cebrian, and I. Rahwan. 2019. "The Evolution of Citation Graphs in Artificial Intelligence Research." *Nature Machine Intelligence* 1 (2): 79–85. doi:10.1038/s42256-019-0024-5

Fromhold-Eisebith, M., P. Marschall, R. Peters, and P. Thomes. 2021. "Torn Between Digitized Future and Context Dependent Past – How Implementing 'Industry 4.0' Production Technologies Could Transform the German Textile Industry." *Technological Forecasting and Social Change* 166: 120620. doi:10.1016/j.techfore.2021.120620

Fröhlich, K., and R. Hassink. 2018. "Regional Resilience: A Stretched Concept?" *European Planning Studies* 26 (9): 1763–1778. doi:10.1080/09654313.2018.1494137

Furman, J., and R. Seamans. 2019. "AI and the Economy." *Innovation Policy and the Economy* 19 (1): 161–191. doi:10.1086/699936

Gentili, A., F. Compagnucci, M. Gallegati, and E. Valentini. 2020. "Are Machines Stealing our Jobs?" *Cambridge Journal of Regions, Economy and Society* 13 (1): 153–173. doi:10.1093/cjres/rsz025

Girasa, R. 2020. *Artificial Intelligence as a Disruptive Technology: Economic Transformation and Government Regulation*. Cham: Springer Nature.

Goel, R. K., C. S. Yadav, and S. Vishnoi. 2021. "Self-sustainable Smart Cities: Socio-Spatial Society Using Participative Bottom-up and Cognitive top-Down Approach." *Cities* 118: 103370. doi:10.1016/j.cities.2021.103370

Gupta, A., P. Panagiotopoulos, and F. Bowen. 2020. "An Orchestration Approach to Smart City Data Ecosystems." *Technological Forecasting and Social Change* 153: 119929. doi:10.1016/j.techfore.2020.119929

Haefner, N., J. Wincent, V. Parida, and O. Gassmann. 2021. "Artificial Intelligence and Innovation Management: A Review, Framework, and Research Agenda?." *Technological Forecasting and Social Change* 162: 120392. doi:10.1016/j.techfore.2020.120392

He, Q., J. Zhou, S. Tan, Y. Song, L. Zhang, Y. Mou, and J. Wu. 2020. "What is the Developmental Level of Outlying Expansion Patches? A Study of 275 Chinese Cities Using Geographical big Data." *Cities* 105: 102395. doi:10.1016/j.cities.2019.102395

Henry, M., D. Schraven, N. Bocken, K. Frenken, M. Hekkert, and J. Kirchherr. 2021. "The Battle of the Buzzwords: A Comparative Review of the Circular Economy and the Sharing Economy Concepts." *Environmental Innovation and Societal Transitions* 38: 1–21. doi:10.1016/j.eist.2020.10.008

Hervás-Oliver, J. L. 2021. "Industry 4.0 in Industrial Districts: Regional Innovation Policy for the Toy Valley District in Spain." *Regional Studies* 55 (10-11): 1775–1786. doi:10.1080/00343404.2021.1939861

Huang, T. L., and B. S. Liu. 2021. "Augmented Reality is Human-Like: How the Humanizing Experience Inspires Destination Brand Love." *Technological Forecasting and Social Change* 170: 120853. doi:10.1016/j.techfore.2021.120853

Hui, E. C., X. Li, T. Chen, and W. Lang. 2020. "Deciphering the Spatial Structure of China's Megacity Region: A new bay Area—The Guangdong-Hong Kong-Macao Greater Bay Area in the Making." *Cities* 105: 102168. doi:10.1016/j.cities.2018.10.011

Indaco, A. 2020. "From Twitter to GDP: Estimating Economic Activity from Social Media." *Regional Science and Urban Economics* 85: 103591. doi:10.1016/j.regsciurbeco.2020.103591

Iqbal, R., F. Doctor, B. More, S. Mahmud, and U. Yousuf. 2020. "Big Data Analytics: Computational Intelligence Techniques and Application Areas." *Technological Forecasting and Social Change* 153: 119253. doi:10.1016/j.techfore.2018.03.024

Jeannerat, H., and T. Theurillat. 2021. "Old Industrial Spaces Challenged by Platformized Value-Capture 4.0." *Regional Studies* 55 (10-11): 1738–1750. doi:10.1080/00343404.2021.1935840

Julsrud, T. E., and J. R. Krogstad. 2020. "Is There Enough Trust for the Smart City? Exploring Acceptance for use of Mobile Phone Data in Oslo and Tallinn." *Technological Forecasting and Social Change* 161: 120314. doi:10.1016/j.techfore.2020.120314

Kagermann, H., W. Wahlster, and J. Helbig. 2013. Recommendations for Implementing the Strategic Initiative INDUSTRIE 4.0. Industrie 4.0 Working Group of Acatech.

Kahle, J. H., É Marcon, A. Ghezzi, and A. G. Frank. 2020. "Smart Products Value Creation in SMEs Innovation Ecosystems." *Technological Forecasting and Social Change* 156: 120024. doi:10.1016/j.techfore.2020.120024

Kandt, J., and M. Batty. 2021. "Smart Cities, big Data and Urban Policy: Towards Urban Analytics for the Long run." *Cities* 109: 102992. doi:10.1016/j.cities.2020.102992

Keding, C. 2021. "Understanding the Interplay of Artificial Intelligence and Strategic Management: Four Decades of Research in Review." *Management Review Quarterly* 71 (1): 91–134. doi:10.1007/s11301-020-00181-x

Kim, B., M. Yoo, K. C. Park, K. R. Lee, and J. H. Kim. 2021. "A Value of Civic Voices for Smart City: A big Data Analysis of Civic Queries Posed by Seoul Citizens." *Cities* 108: 102941. doi:10.1016/j.cities.2020.102941

Kumar, H., M. K. Singh, M. P. Gupta, and J. Madaan. 2020. "Moving Towards Smart Cities: Solutions That Lead to the Smart City Transformation Framework." *Technological Forecasting and Social Change* 153: 119281. doi:10.1016/j.techfore.2018.04.024

Kummitha, R. K. R. 2020. "Why Distance Matters: The Relatedness Between Technology Development and its Appropriation in Smart Cities." *Technological Forecasting and Social Change* 157: 120087. doi:10.1016/j.techfore.2020.120087

Labory, S., and P. Bianchi. 2021. "Regional Industrial Policy in Times of big Disruption: Building Dynamic Capabilities in Regions." *Regional Studies* 55 (10-11): 1829–1838. doi:10.1080/00343404.2021.1928043

Larson, W., and Z. Weihua. 2020. "Self-driving Cars and the City: Effects on Sprawl, Energy Consumption, and Housing Affordability." *Regional Science and Urban Economics* 81. doi:10.1016/j.regsciurbeco.2019.103484

Lazzeretti, L. 2020. "What is the Role of Culture Facing the Digital Revolution Challenge? Some Reflections for a Research Agenda." *European Planning Studies*, 1–21. doi:10.1080/09654313.2020.1836133.

Lepore, D., and F. Spigarelli. 2020. "Integrating Industry 4.0 Plans Into Regional Innovation Strategies." *Local Economy: The Journal of the Local Economy Policy Unit* 35 (5): 496–510. doi:10.1177/0269094220937452

Lim, C., G. H. Cho, and J. Kim. 2021. "Understanding the Linkages of Smart-City Technologies and Applications: Key Lessons from a Text Mining Approach and a Call for Future Research." *Technological Forecasting and Social Change* 170: 120893. doi:10.1016/j.techfore.2021.120893

Liu, W., W. Wu, P. Thakuriah, and J. Wang. 2020. "The Geography of Human Activity and Land use: A big Data Approach." *Cities* 97: 102523. doi:10.1016/j.cities.2019.102523

Ma, Y., J. Wright, S. Gopal, and N. Phillips. 2020. "Seeing the Invisible: From Imagined to Virtual Urban Landscapes." *Cities* 98: 102559. doi:10.1016/j.cities.2019.102559

McAfee, A., and E. Brynjolfsson. (2012). "Big Data: The Management Revolution." *Harvard Business Review* 90 (10): 60–66, 68, 128. PMID: 23074865.

Mohamed, N., J. Al-Jaroodi, I. Jawhar, A. Idries, and F. Mohammed. 2020. "Unmanned Aerial Vehicles Applications in Future Smart Cities." *Technological Forecasting and Social Change* 153: 119293. doi:10.1016/j.techfore.2018.05.004

Monachesi, P. 2020. "Shaping an Alternative Smart City Discourse Through Twitter: Amsterdam and the Role of Creative Migrants." *Cities* 100: 102664. doi:10.1016/j.cities.2020.102664

Olson, A. W., F. Calderón-Figueroa, O. Bidian, D. Silver, and S. Sanner. 2021. "Reading the City Through its Neighbourhoods: Deep Text Embeddings of Yelp Reviews as a Basis for Determining Similarity and Change." *Cities* 110: 103045. doi:10.1016/j.cities.2020.103045

Pasquale, F. 2015. *The Black Box Society: The Secret Algorithms That Control Money and Information*. Cambridge, MA: Harvard University Press.

Perez, C. 2002. *Technological Revolutions and Financial Capital. The Dynamic of Bubbles and Golden Ages*. Cheltenham, UK: Edward Elgar.

Ranchod, R. 2020. "The Data-Technology Nexus in South African Secondary Cities: The Challenges to Smart Governance." *Urban Studies* 57 (16): 3281–3298. doi:10.1177/0042098019896974

Rice, J., and N. Martin. 2020. "Smart Infrastructure Technologies: Crowdsourcing Future Development and Benefits for Australian Communities." *Technological Forecasting and Social Change* 153: 119256. doi:10.1016/j.techfore.2018.03.027

Russell, S., D. Dewey, and M. Tegmark. 2015. "Research Priorities for Robust and Beneficial Artificial Intelligence." *Ai Magazine* 36 (4): 105–114. doi:10.1609/aimag.v36i4.2577

Russell, S., and P. Norvig. 2003. *Artificial Intelligence:A Modern Approach*. 2nd edn. Upper Saddle River, NJ: Prentice Hall.

Sandulli, F. D., E. M. Gimenez-Fernandez, and M. I. Rodriguez Ferradas. 2021. "The Transition of Regional Innovation Systems to Industry 4.0: The Case of Basque Country and Catalonia." *European Planning Studies* 29 (9): 1622–1636. doi:10.1080/09654313.2021.1963049

Sedita, S. R., A. Caloffi, and L. Lazzeretti. 2020. "The Invisible College of Cluster Research: A Bibliometric Core–Periphery Analysis of the Literature." *Industry and Innovation* 27 (5): 562–584. doi:10.1080/13662716.2018.1538872

Shamsuzzoha, A., J. Nieminen, S. Piya, and K. Rutledge. 2021. "Smart City for Sustainable Environment: A Comparison of Participatory Strategies from Helsinki, Singapore and London." *Cities* 114. doi:10.1016/j.cities.2021.103194

Siltori, P. F., R. Anholon, I. S. Rampasso, O. L. Quelhas, L. A. Santa-Eulalia, and W. Leal Filho. 2021. "Industry 4.0 and Corporate Sustainability: An Exploratory Analysis of Possible Impacts in the Brazilian Context." *Technological Forecasting and Social Change* 167: 120741. doi:10.1016/j.techfore.2021.120741

Staletić, N., A. Labus, Z. Bogdanović, M. Despotović-Zrakić, and B. Radenković. 2020. "Citizens' Readiness to Crowdsource Smart City Services: A Developing Country Perspective." *Cities* 107: 102883. doi:10.1016/j.cities.2020.102883

Wamba, S. F., R. E. Bawack, C. Guthrie, M. M. Queiroz, and K. D. A. Carillo. 2021. "Are we Preparing for a Good AI Society? A Bibliometric Review and Research Agenda." *Technological Forecasting and Social Change* 164: 120482. doi:10.1016/j.techfore.2020.120482

Wang, M., and F. Vermeulen. 2021. "Life Between Buildings from a Street View Image: What do big Data Analytics Reveal About Neighbourhood Organisational Vitality?" *Urban Studies* 58 (15): 3118–3139. doi:10.1177/0042098020957198

While, A. H., S. Marvin, and M. Kovacic. 2021. "Urban Robotic Experimentation: San Francisco, Tokyo and Dubai." *Urban Studies* 58 (4): 769–786. doi:10.1177/0042098020917790

Wu, J., Z. Feng, X. Zhang, Y. Xu, and J. Peng. 2020. "Delineating Urban Hinterland Boundaries in the Pearl River Delta: An Approach Integrating Toponym co-Occurrence with Field Strength Model." *Cities* 96: 102457. doi:10.1016/j.cities.2019.102457

Xiao, J., and R. Boschma. 2021. "The Emergence of Artificial Intelligence in European Regions: the Role of a Local ICT Base." Papers in Evolutionary Economic Geography (PEEG) 2117, Utrecht University, Department of Human Geography and Spatial Planning, Group Economic Geography.

Yan, J., J. Liu, and F. M. Tseng. 2020. "An Evaluation System Based on the Self-Organizing System Framework of Smart Cities: A Case Study of Smart Transportation Systems in China." *Technological Forecasting and Social Change* 153: 119371. doi:10.1016/j.techfore.2018.07.009

Yang, Z., D. Wu, and D. Wang. 2021. "Exploring Spatial Path Dependence in Industrial Space with big Data: A Case Study of Beijing." *Cities* 108: 102975. doi:10.1016/j.cities.2020.102975

Zhu, S., W. Jin, and C. He. 2019. 'On Evolutionary Economic Geography: A Literature Review Using Bibliometric Analysis.' *European Planning Studies* 27 (4): 639–660. doi:10.1080/09654313.2019.1568395

Artificial intelligence in hospitality and tourism. State of the art and future research avenues

Martina Nannelli, Francesco Capone and Luciana Lazzeretti

ABSTRACT

The hospitality and tourism industry has recently undergone a profound transformation driven by information and communication technologies developments thanks to innovations in Artificial Intelligence and its tools. However, although the literature on AI and tourism is growing, scientific research on these themes is still fuzzy and fragmented. This article aims to explore the current state of the art and possible future developments of AI and its tools in tourism. The study develops a bibliometric analysis using the ISI database and applies social network analysis to map its intellectual structure. Then, the study develops a qualitative literature review to investigate the main research themes and propose future research avenues. The findings identify some main research strands such as: Big Data for demand forecasting and customer satisfaction; Augmented Reality and Virtual Reality for co-creation processes; the COVID-19 pandemic, social distances issues and service robots; and the smart tourism trends.

1. Introduction

Over the past two decades, the hospitality and tourism (H&T) industry has undergone a profound transformation driven by information and communication technologies (ICTs) that disrupted structures, processes and practices (Gretzel et al. 2015; Sigala 2018). The development of Artificial Intelligence (AI) and its tools is among the main reasons for the current expansion and enhancement of the industry, especially in the field of robotic, virtual reality (VR) and augmented reality (AR) and conversational software. While access to Big Data (BD) sources in the H&T industry is expanding the network of information for increasingly accurate demand forecasting and consumer profiling (Buhalis 2020).

Furman and Seamans (2019, 186) define AI as a '*loose term used to describe a range of advanced technologies that exhibit human-like intelligence including machine learning, autonomous robotics and vehicles, computer vision, language processing, virtual agents, and neural networks*'.

Due to the COVID-19 pandemic, the H&T industry has suffered a setback, making it necessary to rethink the industry to promote a post-pandemic restart in which AI technologies can be the engine for a new renaissance of the whole industry (Hall, Scott, and Gössling 2020).

Studies on AI and its tools are increasing in many fields and disciplines (Lazzeretti et al. 2022). Although the H&T industry is making extensive use of AI tools, studies on the impact of AI and its tools in tourism are still few and mainly application-oriented (Mariani 2020; Doborjeh et al. 2022; Mariani and Baggio 2021). Scholars have pointed out the partiality or lack of theoretical studies on the coupling between tourism and AI technologies (Kirtil and Aşkun 2021; Lv, Shi, and Gursoy 2021), especially with regard to the side effects could be raised (Knani, Echchakoui, and Ladhari 2022).

Therefore, this study explores the current state of the art and possible future developments in AI and its tools in the field of H&T. We combine different methodologies to achieve our aim. The study follows a bibliometric approach, already successfully applied in tourism disciplines (Kirtil and Aşkun 2021; Mariani and Baggio 2021; Capone 2016; Knani, Echchakoui, and Ladhari 2022), to collect data and analyze the spread and distribution of scientific publications, together with a qualitative review of the literature. We collect papers from 1986 to 2021 using ISI Web of Science database (WOS) by combining keywords that identify AI with tourism topics chosen through a preliminary literature review. We then apply Social Network Analysis (SNA) to represent the keywords co-occurrence in the articles to map the intellectual structure of AI in tourism.

Finally, we select the most important articles to conduct a critical literature review to understand the main themes and trends emerging from the topic which allows us to elaborate and discuss new concepts, laying the foundations for a future research agenda.

The study shows that the scientific publications on AI in tourism have drastically increased in the last few years, meaning that AI is becoming more and more central in the field. Moreover, the study identifies and discusses the main emerging topics, namely: the use of BD for demand forecasting and customer satisfaction measurement; the AR and VR experience for value co-creation process; the COVID-19 pandemic and the connection with service robots; finally, the smart tourism trends.

The paper concludes by contributing to the debate on the impact and challenges generated by AI adoption in H&T firms. In fact, although there are several contributions that investigate the AI opportunities for enterprises, there are few papers that attempt to highlight the critical issues and future challenges for tourism and hospitality enterprises.

The article is structured as follows. Section 2 provides an overview of the research on AI and its tools in tourism. Section 3 explains the methodology. Section 4 presents the results of the analysis. Section 5 discusses the relevant topics emerging from the literature review. Finally, conclusions draw suggestions for future research and managerial implications.

2. The previous research on AI in the tourism literature

Technological development has driven the advancement of many service industries reshaping and disrupting the source of innovation and competitive advantage, and theoretical and empirical research in this field (Huang and Rust 2018; Mariani, Perez-Vega, and Wirtz 2022). Among others, the H&T industry has experienced many innovative transformations over time: forms of global distribution systems (Buhalis and Licata 2002), booking and reservation systems (Ghose, Ipeirotis, and Li 2012), social media and mobile applications (Sigala 2018). In particular, what has given the definite push towards technological drift is the developments of AI innovations (Furman and Seamans 2019): new search engines and web data mining (Xiang et al. 2015); new tools to enhance the tourist's experience such as AR and VR devices (Guttentag 2010); new tools for processing ever larger amounts of data such as Machine Learning, Deep Learning, Neural networks and enabling increasingly accurate estimates and forecasts in real-time (Buhalis and Sinatra 2019; Law et al. 2019); new robotics to interact with customers and to perform routine tasks (Belanche et al. 2021).

The COVID-19 pandemic accelerates the use of AI tools in the industry, halting tourist flows and forcing supplementary digitization (Zeng, Chen, and Lew 2020). A common element emerging from several studies and linking the triad of AI, the H&T industry and the pandemic is the increasingly pressing need for accurate estimates and demand forecasting employing advanced computational methodologies (Garcia-Palomares, Gutiérrez, and Mínguez 2015; Tussyadiah and Zach 2017). The application of robotics, such as cleaning and sanitary robots, welcoming services robots and information totems, to tourism represent one of the most relevant innovations. In the same direction, AR and VR tools have also made it possible to enhance the tourist experience, where the difficulties of travelling have made it impossible to experience faraway places (Loureiro, Guerreiro, and Ali 2020).

Some studies have been conducted to trace the empirical development of AI's methods and applications in H&T. For example, Doborjeh et al. (2022) identify AI analytical methods for BD such as Statistical Machine Learning (clustering, classification and prediction techniques), Artificial and Deep Learning Neural Network. In addition, the authors identify both the main current adoptions of AI-based methods (e.g. robotics, AR/VR, chatbots and virtual assistants), and the future methodological and application lines of the industry (e.g. Spatio-Temporal Data analysis, Neuro-tourism and personalized modelling AI applications). Through a systematic review and content analysis, this work specifically focuses on methodological applications in specialized tourism journals, neglecting non-specialized journals and theoretical contributions.

Another interesting work focuses on tourism journals pre-selected upstream of the construction of the database, is that of Lv, Shi, and Gursoy (2021) which have conducted a bibliometric analysis on AI and BD in H&T by reviewing 270 articles published from 2007 to 2020. Findings reveal that, although BD and AI have gradually attracted attention in tourism, attempts are still fuzzy and only a few studies have been conducted to update and summarize the evolution of new technologies in the industry.

As shown by the findings of Mariani and Baggio (2021), who carried out a systematic literature review on BD and analytics, most of the scientific articles are published in academic journals whose main reference areas are not focused on H&T. Therefore, since the subject combines different fields of research in social and hard sciences such as tourism – which is *per se* a multidisciplinary industry – computer science, engineering and mathematics, it seems essential that a comprehensive analysis also considers journals not specializing in tourism.

A special issue on AI in H&T, published in *The International Journal of Contemporary Hospitality and Management* (2021) gathering contributions on specific empirical topics on AI's tools (e.g. robots, chatbots and online reviews), proposes and discusses three lines of research: AI acceptance; AI strength in replacing jobs; AI specific applications. These interesting contributions offer important insights for the industry, especially regarding application and methodological aspects to improve hotel management and guest experience, and to address the current COVID-19 pandemic. Nevertheless, the authors agree on the need for further researches and theoretical studies on this topic, which is a promising but still unexplored area.

Kirtil and Aşkun (2021) carried out a bibliometric analysis searching for the keyword *Artificial Intelligence* in tourism to observe theoretical foundation and progress. The results identify Spain as one of the most published countries and Bournemouth University as the prominent University working on the topic thanks to the several contributions of Buhalis. Moreover, they identify as the main thematic challenges in the sector the lack of understanding and chaotic data management, lack of relevant skills, trust and culture change. Their results differ widely from ours, having searched exclusively for AI. As underlined by the definition of AI here used (Furman and Seamans 2019), we believe, however, that we cannot limit the complex ecosystem that characterizes AI to the use of a single word, but adding other keywords is necessary in order to obtain a 360-degree view on the subject.

An interesting bibliometric study that consolidates the current state of knowledge on AI and tourism is the one proposed by Knani, Echchakoui, and Ladhari (2022) on 1035 peer-reviewed articles in the Scopus and WOS databases. The most important result concerns the evolution of the main research topics that, according to the authors, constitute the issues for a future research agenda and on which this paper attempts to elaborate in the second part of this study: impact of AI on workplaces, privacy concerns, smart tourism experiences, automation and robotics.

These results, on the one hand, highlighted the growing trend of AI research and related topics in the field of tourism, the diffusion of new technological tools throughout the industry and their increasing necessity to overcome several limitations. On the other, they highlighted the need for more studies to support this research context which is still fuzzy and fragmented. Our study attempts to overcome the aforementioned limitations by adopting a multidisciplinary perspective and searching for a larger number of keywords. Furthermore, an important and distinctive element is the in-depth discussion of emerging issues through a qualitative approach.

3. Data source and empirical methodology

The study aims to investigate the current state of the art and future research avenues of AI in H&T using the ISI WOS database and then SNA to explore the intellectual structure of the field through the keywords co-occurrence analysis (Lazzeretti, Sedita, and Caloffi 2014). Finally, a qualitative review of the literature is developed to afford several broad and blurry topics emerging from the previous analysis. Combining these approaches may contribute to better exploring a vast topic that connects multidisciplinary research fields by considering many studies and debating the topic deeply.

This study collects documents from the ISI – WoS database, which is considered the most robust source of quality scientific publications internationally (Sutherland and Jarrahi 2018). We collect papers during the period 1986–2021 by combining keywords closely related to the concept of AI (i.e. big data, machine learning, data mining, etc.) with tourism topics (i.e. tourism, destination, hospitality, visitors, etc.) chosen through a preliminary literature review.[1] The following criteria have been applied to restrict the perimeter of the analysis consistent with the research objective:

- Language: to avoid language constraints, only English contributions have been selected.
- Source: following previous literature, the selection has involved only peer-reviewed articles in scientific journals to ensure the quality of the sample and comparability with other studies.
- Research domain: to maintain a broad perspective on the issue, the selected contributions pertain to the general social sciences categories, such as: Business, Economics, Hospitality Leisure Sport Tourism, Management, Social issues and Urban and Regional studies.

We limited the analysis to articles with at least 1 citation to exclude marginal and weak researches, collecting 1.297 scientific papers. The papers were then manually screened, reading the title, abstract, and, if necessary, the entire document, excluding articles inconsistent with the topics and aim. The final dataset counts 696 articles.

Finally, we combine the previously quantitative methods (SNA) with a qualitative literature review to trace the most discussed topics and help draw an agenda for future research. We identified four most discussed topics combining the results from the SNA keyword co-occurrence analysis and the qualitative literature review. To conduct the qualitative literature review (Section 5), we select the most cited articles based on the higher average citation per year (> 20). We refined our selection by an in-depth reading of abstracts and the entire article when needed. The four topics have been identified carrying out a literature review on the final sample of 63 scientific papers considered relevant to reach our aim.

4. The role of artificial intelligence in tourism

4.1. The evolution of publications on AI in tourism

Figure 1 offers an overview of the evolution of the topic over time. The articles' number on AI in the tourism literature has experienced growth over the years, as expected, reaching up to 166 per year in 2021. This result indicates how hot the topic is and the degree to which the scientific community is investigating it. This growth, however, is very recent. In fact, until 2010, very few articles are found. The first publications started in 1995, and no evidence of scientific papers on the subject has been seen before this date. Simultaneously with the growth in the number of articles, the number of citations also experienced a surge (Figure 2) that follows the general trend of publications. In 2020, citations reached 1.264, almost doubling in just two years. In 2021, Citations decrease in comparison to 2020 as more recent papers usually receive less citations.

Another significant result concerns the journals in which the contributions have been published (Table 1). Most of the articles are published in *Tourism Management* (105) which has more than twice the number of articles published in the second and third journals, namely *International Journal of Contemporary Hospitality Management* (47) and *International Journal of Hospitality Management* (44). These journals are related to

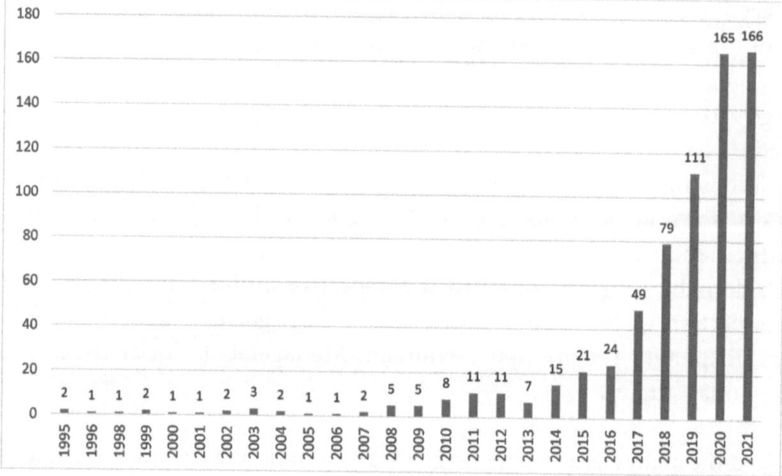

Figure 1. The evolution of publications of AI in Tourism – 1995–2021. *Source: Authors' elaboration based on ISI WoS database*

management disciplines, emphasizing that these issues are primarily discussed in this field of research. In addition, many other journals, among the most published, are related to management and marketing disciplines. Finally, there is a group of journals that, alongside the managerial area, deal with topics about Economics and Social Sciences. Among these, for example, *Current Issues in Tourism* (31), *Annals of Tourist Research* (30), *Tourism Review* (19), etc. Other journals reflect the area of computer science and ICT with management-related topics (*Journal of Hospitality and Tourism Technology* and *Information Technology & Tourism*). Another critical element concerns the entry of journals dedicated to computer-engineering sciences. Among these are *Expert Systems With Applications* (18), *Information Technology & Tourism* (15), *etc.*

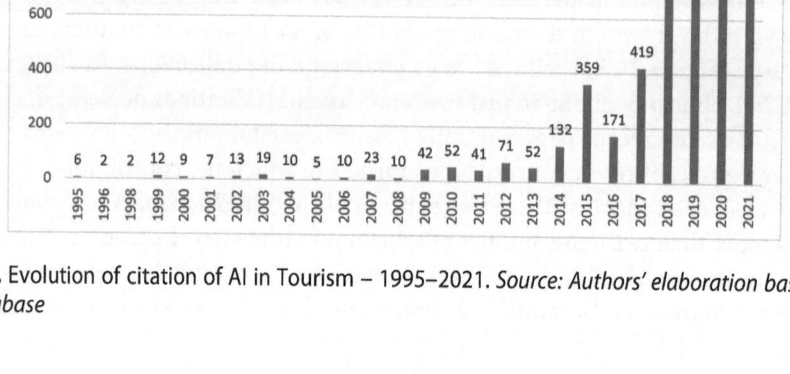

Figure 2. Evolution of citation of AI in Tourism – 1995–2021. *Source: Authors' elaboration based on ISI WoS database*

Table 1. Most publishing journals from 1986 to 2021.

N.	Journal	N. of articles
1	TOURISM MANAGEMENT	105
2	INTERNATIONAL JOURNAL OF CONTEMPORARY HOSPITALITY MANAGEMENT	47
3	INTERNATIONAL JOURNAL OF HOSPITALITY MANAGEMENT	44
4	CURRENT ISSUES IN TOURISM	31
5	ANNALS OF TOURISM RESEARCH	30
6	JOURNAL OF DESTINATION MARKETING & MANAGEMENT	27
7	TOURISM MANAGEMENT PERSPECTIVES	23
8	JOURNAL OF HOSPITALITY AND TOURISM TECHNOLOGY	21
9	JOURNAL OF TRAVEL RESEARCH	19
10	EXPERT SYSTEMS WITH APPLICATIONS	18
11	ISPRS INTERNATIONAL JOURNAL OF GEO-INFORMATION	18
12	JOURNAL OF TRAVEL & TOURISM MARKETING	17
13	TOURISM ECONOMICS	16
14	JOURNAL OF HOSPITALITY MARKETING & MANAGEMENT	16
15	INFORMATION TECHNOLOGY & TOURISM	15
16	TOURISM REVIEW	11
17	JOURNAL OF HOSPITALITY AND TOURISM MANAGEMENT	11
18	INTERNATIONAL JOURNAL OF TOURISM RESEARCH	10
19	ASIA PACIFIC JOURNAL OF TOURISM RESEARCH	10
20	ELECTRONIC MARKETS	8

Source: *Authors' elaboration based on ISI WoS database.*

The search identifies multidisciplinary journals that focus mainly on tourism and new technologies, as expected, given the selection of keywords. However, not choosing the journals a priori allowed us to show that the studied theme is analyzed from different perspectives and disciplines.

4.2. The most publishing authors

There are 1599 authors of the 696 articles for 2138 total co-authorships. The average number of authors per article is 3 confirming the increasing co-authorship trend in many disciplines (Barabási et al. 2002). However, the network of authors is not very cohesive and hierarchical. In fact, out of 1599 authors, 1337 publish only one paper. Only 99 authors publish at least three papers, recording 22% of all papers.

Table 2 shows the top 20 authors with at least six publications on the topic in the period. The table also shows the affiliations at the time of the last publication. These 20 authors alone publish 10% of the total publications. Delving deeper into the table, we can identify important authors such as Law, Gang, Dieck, Mariani and Buhalis at the top and some globally spread schools with much presence in the UK, Australia, China and the USA. Among the most important, we can highlight Manchester Metropolitan University (Dieck, Jung, Jung), Deakin University (Huy, Rong), Bournemouth University (Buhalis), University of Reading (UK) (Mariani and Borghi), etc.

4.3. Mapping the artificial intelligence landscape through keywords co-occurrence

This section investigates and maps the intellectual structure of the field through a keyword co-occurrence analysis. With the help of SNA, it is possible to graphically represent the links among contributions and essential articles' keywords identified with the

Table 2. Most publishing authors.

NR	Author	#	University*	Nationality
1	Law, Rob	38	University of Macau Development Foundation (UMDF)	China
2	Li, Gang	19	School of Hospitality and Tourism Management, Centre for Competitiveness of the Visitor Economy, Surrey	UK
3	Dieck, Claudia Tom	13	Creative AR & VR Hub, Manchester Metropolitan University	UK
4	Mariani, Marcello	13	Henley Business School, University of Reading, Greenlands, Henley on Thames Oxfordshire and University of Bologna	UK
5	Ivanov, Stanislav	10	Varna University of Management, Varna	Bulgaria
6	Buhalis, Dimitrios	9	International Centre for Tourism and Hospitality Research, Bournemouth University, Poole	UK
7	Huy, Quan Vu	9	Department of Information Systems and Business Analytics, Deakin University	Australia
8	Moro, Sergio	8	Instituto Universitário de Lisboa (ISCTE-IUL), ISTAR-IUL, Lisbon	Portugal
9	Song, Haiyan	8	Hospitality and Tourism Research Centre, School of Hotel and Tourism Management, The Hong Kong Polytechnic University, Kowloon	China
10	Jung, Timothy Hyungsoo	7	Creative AR & VR Hub, Manchester Metropolitan University	UK
11	Rita, Paulo	7	NOVA Information Management School (NOVA IMS) Universidade Nova de Lisboa	Portugal
12	Borghi, Matteo	6	Department of Management, University of Reading, Reading	UK
13	Flavian, Carlos	6	Faculty of Economy and Business, University of Zaragoza,	Spain
14	Jung, Timothy	6	Creative AR & VR Hub, Manchester Metropolitan University	UK
15	Kirilenko, Andrei	6	Department of Tourism, Hospitality, and Event Management, University of Florida	USA
16	Pan, Bing	6	Department of Recreation, Park, and Tourism Management, Pennsylvania State University	USA
17	Park, Sangwon	6	Kyung Hee University	South Korea
18	Rong, Jia	6	School of Information Technology, Deakin University	Australia
19	Wang, Shouyang	6	CFS, MDIS, Academy of Mathematics and Systems Science, Chinese Academy of Sciences, Beijing	China
20	Xiang, Zheng	6	The Howard Feiertag Department of Hospitality and Tourism Management, Pamplin College of Business, Virginia Tech, Blacksburg	USA

Source: *our elaboration. Affiliation at the latest articles.*

textual analysis. It was then possible to transform the matrix publication-keywords into a two-mode keywords-keywords graph. Figure 3 represents the network of the most cited keywords (with at least 40 citations) providing information about which research trends (and their connections) are of most concern to researchers.

First of all, it is possible to identify the most frequently cited keywords, effectively replacing a traditional word-cloud graph. Moreover, Figure 3 presents also co-citation, pointing out most relevant interrelationships among keywords. As might be expected, the most cited word is *Big Data*, which has most citations (110 papers cite this word, 16% of the total) showing how BD studies are central to the field of tourism. BD are treated both from a theoretical point of view as a new field of research and empirically for applying new quantitative methodologies. The sub-network of BD is in fact connected to Machine Learning (38 papers) and to other keywords recalling other methodologies like Sentiment Analysis (24), Text Mining (19) and also Social Media (32).

Then the second most cited word is *virtual reality* (53 papers) that together with *augment reality* (40) collect all *Visual Technologies* (93 papers). The keyword *Data mining* is also very cited (44) and connected with online data for tourism analysis through keywords such as *Online review* (41) and *Social Media* (32).

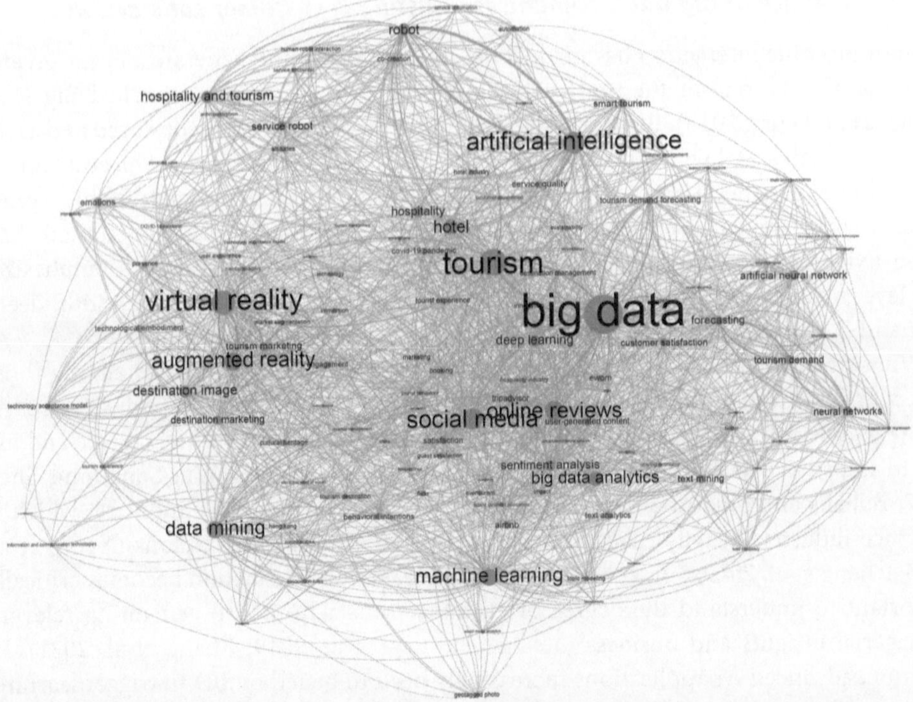

Figure 3. Network of keywords co-occurrence. *Source: Authors' elaboration based on ISI WoS database.*

Artificial intelligence is also very cited (46 papers), but it is not the most cited keyword. AI is connected to *tourism demand forecasting* (22) and *hotel demand forecasting* (21) and *deep learning* (25) algorithms, underlining the role of AI in forecasting demands in tourism.

Robot and *Service Robot* (respectively 31 and 29 papers) accounts together 60 papers, connected in a solid double thread with the word *Artificial Intelligence*. The sub-network is then made up of other related terms (*automation, service automation, human-robot interaction, etc.*), emphasizing and confirming the shift in the hospitality sector towards services robotization and the increasingly humanized robots' characteristics.

An interesting keyword that is worthwhile to cite is *COVID-19* (19 papers). This sub-network confirms tourism destinations' trend towards immersive marketing focusing on direct user experience. This group is crucial because it sets a new industry trend toward the virtual experience. It is also a way to overcome the current pandemic showing how far the emergency has affected tourism.

5. Main research themes of AI in tourism

The second part of the study applies a qualitative approach by analyzing the most cited articles of the literature (>20 average citations per year), by reading the texts in depth to explore some of the emerging themes linking AI to the tourism context. The four topics have been identified carrying out a literature review on the final sample of 63 scientific papers considered relevant to reach our aim. The following sub-sections discuss the 4 topics grouped by thematic interdependence.[2]

5.1. Topic A: use of big data, demand forecasting & customer satisfaction

Human-machine interaction has made it possible to collect and generate an ever-greater amounts of data, one of the most valuable assets of our time, the so-called Big Data (Buhalis and Leung 2018). BD is massive-scale data generated, recorded, stored and accumulated at high speed with advanced technologies and architectures (Lv, Shi, and Gursoy 2021). BD is generated through several sources, including the Internet, mobile transactions, social media, user-generated content and purposefully captured content and so on (Xiang et al. 2015). Moreover, the so-called BD analytics approach emphasizes and leverages the capacity to collect and analyze data with unprecedented width, deepness and scope to address real-life problems (Xiang et al. 2015).

What strongly emerged from the literature is that the H&T industry is among those that contribute most to the generation and use of BD, to the extent the industry ecosystem (e.g. tourist firms, tourism destinations, consumers, etc.) increasingly create and use BD to improve decision-making processes and co-create value (Wu, Song, and Shen 2017; Buhalis and Sinatra 2019; Mariani 2020). Moreover, the H&T industry, which is a service industry, heavily relies on customers feedback and impressions (Xiang et al. 2015; Cheng et al. 2019; Sun et al. 2019). Thus, measuring satisfaction becomes critically important to understand the degree of customer satisfaction, also generating relevant managerial insights and business intelligence (Law et al. 2019; Zhang et al. 2021). In addition, advanced AI applications increasingly need to be fed by BD to trigger learning processes that machines can use to learn from customer behaviours to anticipate and identify future tourists' needs and engage with travellers (Liu et al. 2017).

Many scholars are increasingly interested in adopting BD to understand complex tourism phenomena, also concerning shocks such as the COVID-19 pandemic. Indeed, BD increasingly contributes to developing of the H&T offering the opportunity to enrich research designs and methods as they might be progressively embedded also in mixed methods (Miah et al. 2017). Studies confirm that using BD as a methodology can improve the accuracy of estimates compared to other methods such as questionnaires and interviews, which can also be more costly in terms of time and money (Alaei, Becken, and Stantic 2019). Compared to established techniques, they allow a potentially infinite amount of information to be analyzed, guaranteeing greater comparability of results and improving the generalizability of research findings across different institutional, economic, social and geographical contexts.

Buhalis and Leung (2018) identify two types of BD in the contemporary management of tourist flows: first, internal data such as hotel and guest booking history for smart hospitality; secondly, external contextual information such as economic, political and environmental data and profiles of neighbouring events to conduct a comprehensive business analysis. The strategic combination, elaboration and management of these kinds of data allow the creation of a smart and connected tourism environment operating in real-time (Buhalis and Sinatra 2019).

Nevertheless, despite the prominence of the topic, research on BD in tourism is still somewhat fragmented and predominantly empirical. In fact, all the articles reviewed use BDs as a methodology, but few attempt a more theoretical approach, perhaps due to the very nature of the topic. It is necessary to improve both qualitative and quantitative research to understand how BD can further contribute to the advancement of the industry and develop reading and processing methodologies to provide increasingly accurate, comparable and verifiable estimates (Wu, Song, and Shen 2017; Alaei, Becken, and Stantic 2019).

An important element about BD is the existence of *black holes* to be filled with new research, both theoretical and empirical. Xiang et al. (2015) clarify that theory should play a central role in the analysis of big data, as domain knowledge has proven to be crucial in guiding the data processing and analysis process before reaching the point where meaningful relationships emerge. While Buhalis and Leung (2018) point out some critical elements to overcome: data ownership, accessibility and bias, which may raise numerous questions for the H&T firms.

5.2. Topic B: augmented and virtual reality experience & value co-creation

In the 80s, Jaron Lanier firstly coined the expression virtual reality, an immersive synthetic 3D simulated environment that allows consumers to have the feeling of being in a real-world setting (telepresence) (Vince 2004; Loureiro, Guerreiro, and Ali 2020). The 1990s witnessed the development of VR over the Internet and the use of equipment mainly devoted to gamification. Nevertheless, the books *Introduction to virtual reality* (Vince 2004) and *Stepping into virtual reality* are considered to be the pillars of the modern conception of VR and AR, in the field of tourism only comes to the fore with the article of Guttentag (2010) *Virtual reality: Applications and implications for tourism*, recognized by many authors to be the pillar in the field (Tussyadiah et al. 2018; Loureiro, Guerreiro, and Ali 2020).

VR is widely defined by Guttentag (2010) as '*the use of a computer-generated 3D environment – called a 'virtual environment' – that one can navigate and possibly interact with, resulting in real-time simulation of one or more of the user's five senses*' (638). At the same time, AR is defined as a type of VR.

Of the main elements constituting the VR environment, the *level of immersion* is the factor that can most influence the user's *feelings of presence*, which are, on the one hand, subjective and naturally associated with the user's psychology and, on the other, by the ability of AR systems to deliver high quality (Guttentag 2010; Tussyadiah et al. 2018). Therefore, it links a strong subjective component of the user in receiving high quality data with sensory simulation tools such as input devices that interpret the user's actions and respond contextually. A wide variety of Virtual Environment (VE) can be created and found such as Second Life, where user-avatars can explore VE through social interactions, entertainment and commerce (Guttentag 2010; Tussyadiah et al. 2018). If in the field of tourism, the first application was found in the project Rome Reborn, a 3D virtual model of ancient Rome, which uses virtual sites' reconstructions and 3D edutainment through avatars (Guttentag 2010).

The first area of application of AR in tourism is in firms planning and management, where it can have several implications. First, the power of VR to aid in strategy formulation through the power of being able to virtually observe an environment and its changes, and to easily communicate to a community through direct immersion. Secondly, in marketing due to VR high capacity to sponsor a destination, tourism products or services by attracting tourists and consumers remotely (Bogicevic et al. 2019; Kim, Lee, and Jung 2020), and to communicate with a higher degree of trust (Buhalis and Law 2008;

Buhalis and Sinatra 2019). The third area of application concerns entertainment. The gaming field is one of the first to emerge in connection with VR themes and one of the most promising. Although many entertainment applications in VR are designed for home use, others have already established themselves as attractions in tourist destinations (e.g. theme parks) (Jung, Chung, and Leue 2015; Wei, Qi, and Zhang 2019). Education is linked to this fourth field of application, as AR is a powerful interactive tool for teaching and learning directly from the 'synthetic' experience (Guttentag 2010; Kim, Lee, and Jung 2020). Finally, several researchers have suggested that VR could help to preserve cultural heritage by providing an alternative form of access to fragile sites and environments (Guttentag 2010; Tussyadiah et al. 2018). VR can protect sites, such as UNESCO World Heritage sites that attract large numbers of visitors, from damage caused by overtourism in fragile environments while preserving the sustainability of places. VR and AR firms can then benefit from this business and opportunities (Lazzeretti and Capone 2016).

Not only VR and related technologies are making a revolution in the way tourists experience travel and tourism-related products and services, but firms may also use AI as powerful tools to co-create value with users challenging the H&T industry itself (Dieck and Jung 2018; Tung and Au 2018) due to the possibility of direct immersion in the tourist experience and of interaction between producer and user, enabling experience environment in which every ecosystem player is highly engaged. Buhalis and Sinatra (2019) identify how brands in tourism utilize digital technology to enhance the consumer experience in real-time to co-create value enabling consumers to co-construct their unique personalized experience (Buhalis et al. 2019) where VR and AR tools act as a medium to engage in real-time (Buhalis and Foerste 2015).

Finally, some scholars have speculated that the Metaverse will have a profound impact on the hospitality and tourism industry in the future (Gursoy, Malodia, and Dhir 2022). Research in this area is in its infancy, but in the industry, along with VR/AR improvements, the Metaverse could have a growing impact on tourism, and at the moment, both academia and industry face multiple critical issues that will be interesting to investigate in the future (Gursoy, Malodia, and Dhir 2022).

5.3. Topic C: COVID-19 pandemic, health care, social distances & service robots

Numerous empirical and theoretical studies have been conducted to address the threats and challenges posed by the COVID-19 pandemic which shutdown the entire tourism industry. Although the authors agree that it is still too early to know what travel and tourism will be like after we finally overcome the pandemic worldwide, scholars agree that the ICTs will continue to play a significant role in various strategic sectors such as transport, hospitality, catering and access to tourist sites and attractions (Jiang and Wen 2020; Fotiadis, Polyzos, and Huan 2021). Studies on the implementation of technology in tourism, in particular, have taken two main directions: on the one hand, studies on the use of BD and AI methodologies for demand forecasting and customer satisfaction, discussed in Topic A; on the other hand, studies on the adoption of robotic technology addressed below.

A conspicuous strand of research on AI and tourism concerns the robotization of the industry in terms of consumer preference and attitude to interface with a human-like robot (de Kervenoael et al. 2020; Pillai et al. 2021). Indeed, robots are often used in response to disasters and shocks, such as the Fukushima nuclear power plant catastrophe (Zeng, Chen, and Lew 2020). However, with advances in AI and other technologies, robotics has become increasingly viable in hospitality firms to provide concierge, cleaning, catering and other services (Cain, Thomas, and Alonso 2019; Jiang and Wen 2020; Yu 2020). In the COVID-19 pandemic, robots initially proved particularly effective for routine or non-routine tasks, such as using ultraviolet light to disinfect surfaces. Later, however, robot technology matured rapidly and different types of robots appeared to handle COVID-19 in various environments, including hospitals, airports, transport, recreational and scenic areas, hotels, and communities. In particular, robots have taken on some essential tasks in protecting the community through disinfecting and monitoring spaces, ensuring social distancing through autonomous transport, delivering necessities within hospitals, airports and stations, and assisting managers in safety, real-time information and directional services.

The technological drift of the industry brings contradictions and consequences for firms in H&T (Lu, Cai, and Gursoy 2019). Seyitoğlu and Ivanov (2021) assert that empathy and emotional intelligence are essential drivers of customer perceptions, service experience and quality (Huang and Rust 2018). However, service robots may increase the psychological distance between tourists and residents in a destination due to their decreased physical interaction. Kim et al. (2021), applying four experiments to research tourists' liking for robots over humans, showed that the pandemic shifted liking towards services provided by robots. Tung and Au (2018) suggest that human-machine interaction can lead to the co-creation of values, and it is, therefore, essential to assess areas where the application of machines can positively influence consumer perceptions. Murphy, Gretzel, and Pesonen (2019) suggest intervening directly on those anthropomorphized machines' features to decrease the distance with customers by addressing the anthropomorphism of machines.

The second argument about service robots concerns their convenience and efficiency compared to humans. Indeed, service robots can reduce operating costs and provide customers with better services. Belanche et al. (2021) found that the customer's affinity with the service robot positively influences the attribution of service improvement, which in turn has a positive influence on the customer's behavioural intentions. Conversely, affinity hurts the attribution of cost reduction, which harms behavioural purposes. Finally, human similarity has a positive influence on affinity.

We are certainly seeing a shift in perception between pre- and post-pandemic. Perhaps, the industry should push towards robotization in those phases where user interaction is low or non-existent such as cleaning services, and focus on human interaction in those services where face-to-face dialogue can enhance the experience, such as catering and hospitality (Huang and Rust 2018; Seyitoğlu and Ivanov 2021).

5.4. Topic D: smart tourism trends

In recent years, a new and global search stream has emerged under the umbrella of the *smart* tag. Smartness has found one of the greatest strengths in conjunction with the context of cities to describe an environment where city elements and technologies interact in innovative ways to achieve resource optimization, effective and equitable governance, sustainability and quality of life: smart city (Gretzel et al. 2015; Adler and Florida 2021). What remains unclear is the definition of the smartness. Smartness is defined as a new *buzzword* to describe technological, economic and social developments supported by technologies that rely on sensors, BD, open data, new modes of connectivity and information (e.g. Internet of Things; Internet of Everything), and the ability to infer and reason through complex methodologies such as deep learning, machine learning (Gretzel et al. 2015). If *smart* is used to describe a complex mixture of several interconnected elements, certainly the technological part holds the central position. Both Buhalis and Leung (2018) and Adler and Florida (2021) state that the concept of smartness refers to a network of firms, organizations and smart functions that engage in interoperable and interconnected systems to simplify and automate everyday activities and (co-) create value across the ecosystem for all stakeholders.

In the H&T industry the term *smart tourism*, as well as *smart destination* or *smart tourism experience*, is becoming more and more fashionable, as shown by the size of the word *smart* in the keyword co-occurrence analysis (section 4.3). Many smart tourism initiatives have arisen from smart city projects and, consequently, from smart tourist destinations.

Gretzel et al. (2015) identify three main levels of smart tourism: smart destinations, i.e. exceptional cases of smart cities that consider not only residents, but also tourists in dealing with mobility, resource supply and distribution, sustainability and high-quality of life/visits; *smart experience* that focuses particularly on technology-mediated tourism experiences and their enhancement through personalization, context awareness and real-time monitoring, and where users contribute data through photos, feedback, etc.; finally, the third layer is the *smart business* which refers to the complex business ecosystem that creates and supports the exchange of touristic resources and the co-creation of the tourism experience. Adler and Florida (2021) focus their study on smart tourism experience, which should include smart technologies and solutions to offer data-driven, built-in real-time, context-aware and co-created experiences.

Buhalis and Leung (2018) focus on the smart hospitality ecosystem, which involves many direct and indirect stakeholders, proposing an innovative framework at the heart of which two main elements are identified: interoperability and interconnectivity. Indeed, smartness can effectively develop networks to create an ecosystem and dynamically interconnect all members. Interoperability refers to the ability of business systems and processes to support the exchange of data and enable the spread of information and knowledge. Interconnectivity minimizes barriers to collaboration, thus helping hotels to constantly assess needs and strategies. Although the focus is on hospitality, the authors state that management can easily apply technology to support tourism '*as a highly-interconnected and networked industry*' (Buhalis and Leung 2018, 42) and empower the co-creation process.

One of the main problems in smart tourism is the lack of (big) data organization, standardization and data access. Furthermore, Gretzel et al. (2015) state that tourism is still not a sector that attracts many knowledge workers, and despite its heavy reliance on ICT, it is known to have problems with a lack of innovation.

The development of smart tourism is certainly already underway due to the widespread deployment of technology in the industry. However, systematic coordination and the exploitation of data sources such as BD for value creation are still in their infancy. Although smart tourism initiatives worldwide promote the construction of viable smart tourism ecosystems (Gretzel et al. 2015), the complexity of the industry makes it very difficult to go beyond specific innovations. Therefore, further research needs to be conducted to clarify the meaning of the term and to define the elements that make up the ecosystem for smart tourism, and empirical studies need to be undertaken to explore the ecosystem through new statistical methodologies.

6. Conclusions

The present research aims to investigate the development of AI in scientific research in H&T. The work developed a joint bibliometric analysis with an analysis of keywords co-occurrence, along with a qualitative analysis of the literature to approach the topic from different perspectives.

While there have been a couple of literature reviews covering BD in H&T up to 2017 (Li et al. 2017; Mariani et al. 2018), to the best of our knowledge, there are only few studies on AI in tourism research (Kirtil and Aşkun 2021; Lv, Shi, and Gursoy 2021; Knani, Echchakoui, and Ladhari 2022). This study thus provides further insights into the evolution of research and presents an up-to-date overview.

Our first result confirms Mariani's (2020) and Lv, Shi, and Gursoy (2021) research about the exponential growth of studies on AI and tourism in recent years. Thus, our focus has pointed out that the journals involved in the debate are not only specialists on tourism topics, and tourism has undergone a disciplinary broadening in this regard (Mariani and Baggio 2021). However, in recent years this phenomenon is changing, and AI and tourism are also registering an increasing centrality in tourism journals. Like other authors, we have highlighted the presence of many contributions that use AI and BD exclusively as tools for tourism research emphasizing the lack of more purely theoretical research. However, exclusive fields in which AI and tourism can be developed were also highlighted. Indeed, we discussed the opportunities and challenges that COVID19 has created and how they can be addressed through AI applications such as AR, VR and robots.

The interdisciplinarity of the topic is also confirmed by the network of authors. Indeed, although some networks of collaborating authors have been identified, the study shows that networks of collaborations are still not particularly intense, perhaps also due to the youth of the topic itself.

An interesting result concerns the topics most discussed in the scientific literature, which would seem to constitute the directions in which the H&T industry is moving. Among these, four main thematic blocks have been identified: the first revolves around the role of BD for demand prediction and consumer care; the second concerns the use of new tools aimed at improving the consumer experience, AR and VR; the third group looks at the recent COVID-19 pandemic and the possibilities for the industry, particularly in the field of robotics; finally, the fourth concerns the theme of smartness applied to the H&T industry with a perspective of developing an increasingly interconnected industry.

In particular, AI-based tools have advanced tourism research, enabling the use of BD, which is particularly useful for firms and managers in demand forecasting, online customer satisfaction and all studies of social networks and online data. However, this poses challenges to firms, tourism organizations and institutions in integrating business intelligence (Mariani 2020) and developing expertise in these tools.

The use of AI in tourism offers firms many opportunities, but it also needs to be accompanied by better training either at work level or at university/school level in developing teams with more analytical and multidisciplinary skills, otherwise firms and managers will struggle with new trends in industry development and data analysis.

It is not all positive; however, there is also a dark side of AI in tourism research. Lazzeretti (2020) highlights the problems related to the nature of the algorithms, the *black box* principles of their operation, as well as the use of data *in the wild*, biases and hidden errors. Furthermore, the use of data created for non-specific purposes and non-transparent tools (random forest or neural network) hinder an even wider dissemination of these tools.

Indeed, firms in H&T and the whole eco-systems can also benefit significantly from new technologies and the use of AI and BD as a service industry, very much oriented to online platforms, tourist reviews, customer satisfaction and the need to forecast demand and predict consumer behaviour. However, like other disciplines, it faces some challenges and dangers that must be addressed.

Another challenge emerges from the difficulties for firms to adopt AI in their activities and tasks. In this paper, we have used a general definition of AI (Furman and Seamans 2019), while some studies indicate that AI can be adopted differently depending on the type of task being considered. Huang and Rust (2018) specify four ordinal and parallel types of intelligence – mechanical, analytical, intuitive and empathetic – in the order of difficulty with which AI masters them. For example, replacing simple and repetitive tasks involves 'mechanical intelligence' and it is then easier for firms to adopt AI to replace these tasks.[3] In services, and in particular in H&T, jobs and tasks however involve more 'intuitive' and 'empathetic' intelligence and so it could be more difficult for firms to replace a whole range of tasks that will continue to be handled by people. For instance, hotels hospitality staff can be replaced by virtual bots, but this would lead to totally change the consumer experience (Huang and Rust 2018).

Finally, it is possible to reflect on some managerial implications of the work. We have highlighted how AI can offer several opportunities for companies and managers working in H&T. The opportunities are disparate ranging from data mining, big data analytics, demand forecasting with machine learning technologies, and innovative service delivery with AR/VR, etc. It would be reductive to propose an exhaustive list, however it is important to emphasize that we have also highlighted difficulties and critical issues. A first problem concerns the availability of skills and competencies of the human capital that will have to use and manage the new technologies. This will certainly involve investments, including in technology, and will surely raise challenges mainly to small and medium-sized H&T enterprises. Future technological developments will be crucial, although there is no shortage of challenges and difficulties in this technological change.

In conclusion, this article has provided an up-to-date overview of AI research in scientific research in tourism, describing the main trends of study, the most influential authors, the most used journals and the most developed techniques. We are at a very relevant historical moment, AI and BD methodologies are no longer borrowed from other fields and occasionally applied to tourism but are increasingly developed and central to published works and studies. The future is expected to confirm this trend, and scholars, firms and managers will have to equip themselves with skills and capabilities that are increasingly needed in a society that is constantly changing technologically, and tourism cannot stand aside.

Notes

1. The selected keywords are: 'artificial intelligence', 'big data', 'data mining', 'machine learning', 'deep learning', 'neural network', 'robot' and 'accommodation', 'destination', 'holiday', 'hospitality', 'hotel', 'journey', 'tourism', 'travel', 'vacation', 'visit'. All in single and plural forms. See for instance other work that use similar keywords (Lv, Shi, and Gursoy 2021; Lazzeretti et al. 2022, etc.).
2. The co-occurrence analysis developed in Section 4.3 identifies some major sub-networks that partially confirm the qualitative literature review, but it is not coincident. We believe that the two analyzes cannot and should not necessarily coincide in their results. In fact, the quantitative analysis lays the groundwork for the identification of the 4 topics on which the qualitative review is based.
3. For instance Virtual bots that turn customer service into self-service.

Disclosure statement

No potential conflict of interest was reported by the author(s).

Funding

This work was supported by CHANGES, PNRR, Mission 4, PE5, NextGenEU (CUP B53C22004010006).

ORCID

Francesco Capone http://orcid.org/0000-0003-2000-3033
Luciana Lazzeretti http://orcid.org/0000-0002-9759-2289

References

Adler, P., and R. Florida. 2021. "The Rise of Urban Tech: How Innovations for Cities Come from Cities." *Regional Studies* 55 (10-11): 1787–1800. doi:10.1080/00343404.2021.1962520

Alaei, A., S. Becken, and B. Stantic. 2019. "Sentiment Analysis in Tourism: Capitalising on big Data." *Journal of Travel Research* 58 (2): 175–191. doi:10.1177/0047287517747753

Barabási, A., H. Jeong, Z. Neda, E. Ravasz, A. Schubert, and T. Vicsek. 2002. "Evolution of the Social Network of Scientific Collaborations." *Physica A: Statistical Mechanics and its Applications* 311 (3–4): 590–614. doi:10.1016/S0378-4371(02)00736-7

Belanche, D., L. Casaló, J. Schepers, and C. Flavián. 2021. "Examining the Effects of Robots' Physical Appearance, Warmth, and Competence in Frontline Services: The Humanness-Value-Loyalty Model." *Psychology & Marketing* 38 (12): 2357–2376. doi:10.1002/mar.21532

Bogicevic, V., S. Seo, J. Kandampully, S. Liu, and N. Rudd. 2019. "Virtual Reality Presence as a Preamble of Tourism Experience: The Role of Mental Imagery." *Tourism Management* 74: 55–64. doi:10.1016/j.tourman.2019.02.009

Buhalis, D. 2020. "Technology in Tourism-from Information Communication Technologies to ETourism and Smart Tourism Towards Ambient Intelligence Tourism: A Perspective Article." *Tourism Review* 75 (1): 67–272. doi:10.1108/TR-06-2019-0258

Buhalis, D., and M. Foerste. 2015. "SoCoMo Marketing for Travel and Tourism: Empowering co-Creation of Value." *Journal of Destination Marketing & Management* 4 (3): 151–161. doi:10.1016/j.jdmm.2015.04.001

Buhalis, D., T. Harwood, V. Bogicevic, G. Viglia, S. Beldona, and C. Hofacker. 2019. "Technological Disruptions in Services: Lessons from Tourism and Hospitality." *Journal of Service Management.* doi:10.1108/JOSM-12-2018-0398

Buhalis, D., and R. Law. 2008. "Progress in Information Technology and Tourism Management: 20 Years on and 10 Years After the Internet-The State of ETourism Research." *Tourism Management* 29 (4): 609–623. doi:10.1016/j.tourman.2008.01.005

Buhalis, D., and R. Leung. 2018. "Smart Hospitality - Interconnectivity and Interoperability Towards an Ecosystem." *International Journal of Hospitality Management* 71: 41–50. doi:10.1016/j.ijhm.2017.11.011

Buhalis, D., and M. Licata. 2002. "The Future ETourism Intermediaries." *Tourism Management* 23 (3): 207–220. doi:10.1016/S0261-5177(01)00085-1

Buhalis, D., and Y. Sinatra. 2019. "Real-time co-Creation and Nowness Service: Lessons from Tourism and Hospitality." *Journal of Travel and Tourism Marketing* 36 (5): 563–582. doi:10.1080/10548408.2019.1592059

Cain, L., J. Thomas, and M. Alonso Jr. 2019. "From sci-fi to sci-Fact: The State of Robotics and AI in the Hospitality Industry." *Journal of Hospitality and Tourism Technology.* doi:10.1108/JHTT-07-2018-0066

Capone, F. 2016. *Tourist Clusters, Destinations and Competitiveness.* Cheltenham.: Routledge.

Cheng, X., S. Fu, J. Sun, A. Bilgihan, and F. Okumus. 2019. "An Investigation on Online Reviews in Sharing Economy Driven Hospitality Platforms: A Viewpoint of Trust." *Tourism Management* 71: 366–377. doi:10.1016/j.tourman.2018.10.020

de Kervenoael, R., R. Hasan, A. Schwob, and E. Goh. 2020. "Leveraging Human-Robot Interaction in Hospitality Services: Incorporating the Role of Perceived Value, Empathy, and Information Sharing Into Visitors' Intentions to use Social Robots." *Tourism Management* 78: 104042. doi:10.1016/j.tourman.2019.104042

Dieck, M., and T. Jung. 2018. "A Theoretical Model of Mobile Augmented Reality Acceptance in Urban Heritage Tourism." *Current Issues in Tourism* 21 (2): 154–174. doi:10.1080/13683500.2015.1070801

Doborjeh, Z., N. Hemmington, M. Doborjeh, and N. Kasabov. 2022. "Artificial Intelligence: A Systematic Review of Methods and Applications in Hospitality and Tourism." *International Journal of Contemporary Hospitality Management* 34 (3): 1154–1176. doi:10.1108/IJCHM-06-2021-0767

Fotiadis, A., S. Polyzos, and T. Huan. 2021. "The Good, the bad and the Ugly on COVID-19 Tourism Recovery." *Annals of Tourism Research* 87: 103–117. doi:10.1016/j.annals.2020.103117

Furman, J., and R. Seamans. 2019. "AI and the Economy." *Innovation Policy and the Economy* 19 (1): 161–191. doi:10.1086/699936

Garcia-Palomares, J., J. Gutiérrez, and C. Mínguez. 2015. "Identification of Tourist hot Spots Based on Social Networks: A Comparative Analysis of European Metropolises Using Photo-Sharing Services and GIS." *Applied Geography* 63: 408–417. doi:10.1016/j.apgeog.2015.08.002

Ghose, A., P. Ipeirotis, and B. Li. 2012. "Designing Ranking Systems for Hotels on Travel Search Engines by Mining User-Generated and Crowdsourced Content." *Marketing Science* 31 (3): 493–520. doi:10.1287/mksc.1110.0700

Gretzel, U., M. Sigala, Z. Xiang, and C. Koo. 2015. "Smart Tourism: Foundations and Developments." *Electronic Markets* 25 (3): 179–188. doi:10.1007/s12525-015-0196-8

Gursoy, D., S. Malodia, and A. Dhir. 2022. "The Metaverse in the Hospitality and Tourism Industry: An Overview of Current Trends and Future Research Directions." *Journal of Hospitality Marketing & Management*, 1–8. doi:10.1080/19368623.2022.2072504

Guttentag, D. 2010. "Virtual Reality: Applications and Implications for Tourism." *Tourism Management* 31 (5): 637–651. doi:10.1016/j.tourman.2009.07.003

Hall, C., D. Scott, and S. Gössling. 2020. "Pandemics, Transformations and Tourism: Be Careful What you Wish for." *Tourism Geographies* 22 (3): 577–598. doi:10.1080/14616688.2020.1759131

Huang, M., and R. Rust. 2018. "Artificial Intelligence in Service." *Journal of Service Research* 21 (2): 155–172. doi:10.1177/1094670517752459

Jiang, Y., and J. Wen. 2020. "Effects of COVID-19 on Hotel Marketing and Management: A Perspective Article." *International Journal of Contemporary Hospitality Management* 32 (8): 2563–2573. doi:10.1108/IJCHM-03-2020-0237

Jung, T., N. Chung, and M. Leue. 2015. "The Determinants of Recommendations to use Augmented Reality Technologies: The Case of a Korean Theme Park." *Tourism Management* 49: 75–86. doi:10.1016/j.tourman.2015.02.013

Kim, S., J. Kim, F. Badu-Baiden, M. Giroux, and Y. Choi. 2021. "Preference for Robot Service or Human Service in Hotels? Impacts of the COVID-19 Pandemic." *International Journal of Hospitality Management* 93: 102795. doi:10.1016/j.ijhm.2020.102795

Kim, M., C. Lee, and T. Jung. 2020. "Exploring Consumer Behavior in Virtual Reality Tourism Using an Extended Stimulus-Organism-Response Model." *Journal of Travel Research* 59 (1): 69–89. doi:10.1177/0047287518818915

Kirtil, I., and V. Aşkun. 2021. "Artificial Intelligence in Tourism: A Review and Bibliometrics Research." *Advances in Hospitality and Tourism Research (AHTR)* 9 (1): 205–233. doi:10.30519/ahtr.801690

Knani, M., S. Echchakoui, and R. Ladhari. 2022. "Artificial Intelligence in Tourism and Hospitality: Bibliometric Analysis and Research Agenda." *International Journal of Hospitality Management* 107: 103317. doi:10.1016/j.ijhm.2022.103317

Law, R., G. Li, D. Fong, and X. Han. 2019. "Tourism Demand Forecasting: A Deep Learning Approach." *Annals of Tourism Research* 75: 410–423. doi:10.1016/j.annals.2019.01.014

Lazzeretti, L. 2020. "What is the Role of Culture Facing the Digital Revolution Challenge? Some Reflections for a Research Agenda." *European Planning Studies*, 1–21. doi:10.1080/09654313.2020.1836133

Lazzeretti, L., and F. Capone. 2016. "R&D Networks in High Technology Applied to Cultural Goods in Tuscany. A Social Network Analysis." *Sinergie Italian Journal of Management* 34: 63–87. doi:10.7433/s99.2016.06

Lazzeretti, L., N. Innocenti, M. Nannelli, and S. Oliva. 2022. "The Emergence of Artificial Intelligence in the Regional Sciences: A Literature Review." *European Planning Studies*, 1–21. doi:10.1080/09654313.2022.2101880

Lazzeretti, L., S. Sedita, and A. Caloffi. 2014. "Founders and Disseminators of Cluster Research." *Journal of Economic Geography* 14 (1): 21–43. doi:10.1093/jeg/lbs053

Liu, Y., T. Teichert, M. Rossi, H. Li, and F. Hu. 2017. "Big Data for big Insights: Investigating Language-Specific Drivers of Hotel Satisfaction with 412,784 User-Generated Reviews." *Tourism Management* 59: 554–563. doi:10.1016/j.tourman.2016.08.012

Loureiro, S., J. Guerreiro, and F. Ali. 2020. "20 Years of Research on Virtual Reality and Augmented Reality in Tourism Context: A Text-Mining Approach." *Tourism Management* 77: 104028. doi:10.1016/j.tourman.2019.104028

Lu, L., R. Cai, and D. Gursoy. 2019. "Developing and Validating a Service Robot Integration Willingness Scale." *International Journal of Hospitality Management* 80: 36–51. doi:10.1016/j.ijhm.2019.01.005

Lv, H., S. Shi, and D. Gursoy. 2021. "A Look Back and a Leap Forward: A Review and Synthesis of big Data and Artificial Intelligence Literature in Hospitality and Tourism." *Journal of Hospitality Marketing and Management*, 1–31. doi:10.1080/19368623.2021.1937434

Mariani, M. 2020. "Big Data and Analytics in Tourism and Hospitality: A Perspective Article." *Tourism Review* 75: 299–303. doi:10.1108/TR-06-2019-0259

Mariani, M., and R. Baggio. 2021. "Big Data and Analytics in Hospitality and Tourism: A Systematic Literature Review." *International Journal of Contemporary Hospitality Management* 34: 231–278. doi:10.1108/IJCHM-03-2021-0301

Mariani, M., R. Baggio, M. Fuchs, and W. Höepken. 2018. "Business Intelligence and big Data in Hospitality and Tourism: A Systematic Literature Review." *International Journal of Contemporary Hospitality Management* 30 (12): 3514–3554. doi:10.1108/IJCHM-07-2017-0461

Mariani, M., R. Perez-Vega, and J. Wirtz. 2022. "AI in Marketing, Consumer Research and Psychology: A Systematic Literature Review and Research Agenda." *Psychology & Marketing* 39 (4): 755–776. doi:10.1002/mar.21619

Miah, S., H. Vu, J. Gammack, and M. McGrath. 2017. "A big Data Analytics Method for Tourist Behaviour Analysis." *Information & Management* 54: 771–785. doi:10.1016/j.im.2016.11.011

Murphy, J., U. Gretzel, and J. Pesonen. 2019. "Marketing Robot Services in Hospitality and Tourism: The Role of Anthropomorphism." *Future of Tourism Marketing*, 16–27. Routledge. doi:10.1080/10548408.2019.1571983

Pillai, S., K. Haldorai, W. Seo, and W. Kim. 2021. "COVID-19 and Hospitality 5.0: Redefining Hospitality Operations." *International Journal of Hospitality Management* 94: 102869. doi:10.1016/j.ijhm.2021.102869

Seyitoğlu, F., and S. Ivanov. 2021. "Service Robots as a Tool for Physical Distancing in Tourism." *Current Issues in Tourism* 24 (12): 1631–1634. doi:10.1080/13683500.2020.1774518

Sigala, M. 2018. "New Technologies in Tourism: From Multidisciplinary to Anti-Disciplinary Advances and Trajectories." *Tourism Management Perspectives* 25: 151–155. doi:10.1016/j.tmp.2017.12.003

Sun, S., Y. Wei, K. Tsui, and S. Wang. 2019. "Forecasting Tourist Arrivals with Machine Learning and Internet Search Index." *Tourism Management* 70: 1–10. doi:10.1016/j.tourman.2018.07.010

Sutherland, W., and M. Jarrahi. 2018. "The Sharing Economy and Digital Platforms: A Review and Research Agenda." *International Journal of Information Management* 43: 328–341. doi:10.1016/j.ijinfomgt.2018.07.004

Tung, V., and N. Au. 2018. "Exploring Customer Experiences with Robotics in Hospitality." *International Journal of Contemporary Hospitality Management* 30 (7): 2680–2697. doi:10.1108/IJCHM-06-2017-0322

Tussyadiah, I., D. Wang, T. Jung, and M. Tom Dieck. 2018. "Virtual Reality, Presence, and Attitude Change." *Empirical Evidence from Tourism. Tourism Management* 66: 140–154. doi:10.1016/j.tourman.2017.12.003

Tussyadiah, I., and F. Zach. 2017. "Identifying Salient Attributes of Peer-to-Peer Accommodation Experience." *Journal of Travel & Tourism Marketing* 34: 636–652. doi:10.1080/10548408.2016.1209153

Vince, J. 2004. *Introduction to Virtual Reality*. London: Springer Science & Business Media.

Wei, W., R. Qi, and L. Zhang. 2019. "Effects of Virtual Reality on Theme Park Visitors' Experience and Behaviors: A Presence Perspective." *Tourism Management* 71: 282–293. doi:10.1016/j.tourman.2018.10.024

Wu, D., H. Song, and S. Shen. 2017. "New Developments in Tourism and Hotel Demand Modeling and Forecasting." *International Journal of Contemporary Hospitality Management* 29 (1): 507–529. doi:10.1108/IJCHM-05-2015-0249

Xiang, Z., Z. Schwartz, J. Gerdes Jr, and M. Uysal. 2015. "What Can big Data and Text Analytics Tell us About Hotel Guest Experience and Satisfaction?" *International Journal of Hospitality Management* 44: 120–130. doi:10.1016/j.ijhm.2014.10.013

Yu, C. 2020. "Human-like Robots as Employees in the Hotel Industry: Thematic Content Analysis of Online Reviews." *Journal of Hospitality Marketing & Management* 29: 22–38. doi:10.1080/19368623.2019.1592733

Zeng, Z., P. Chen, and A. Lew. 2020. "From High-Touch to High-Tech: COVID-19 Drives Robotics Adoption." *Tourism Geographies* 22: 724–734. doi:10.1080/14616688.2020.1762118

Zhang, Y., G. Li, B. Muskat, and R. Law. 2021. "Tourism Demand Forecasting: A Decomposed Deep Learning Approach." *Journal of Travel Research* 60: 981–997. doi:10.1177/0047287520919522

Learning as imitation or mimesis: how 'smart' is machine learning for its planning controllers?

Philip Cooke

ABSTRACT
The idea to be explored in this contribution is that to understand change as society evolves is useless without Learning that a little knowledge can be a dangerous thing and may look 'smart' but is likely to be unwise. This in turn requires an understanding of why the label 'knowledge economy' came into substantial usage by opinion-formers at about the same time. Thus a first wave of injunctions in favour of 'Learning' by governments and corporate leaders occurred about 30 years ago. The change in question was led by information technology and the production and consumption practices it entailed. In a second wave of 'Learning from Leaders', especially how to be 'Smart', the lesson quickly became 'Learning from Losers'. Here some of the most-vaunted - for example – 'smart' visions for various functions nevertheless failed to deliver. Perhaps the greatest failure to learn has been the sight and sound of 'Flailing by Failing' from Science Policy 'led' governments in response to the SARS2-Covid-19 pandemic when the lessons of at least moderate success involved 'Learning from Life' after being prepared by previous experience.

Introduction

This paper focuses on the rise and gradual demise of the epithet 'learning' as a pre-modifier. The contribution argues, uncontroversially, that its rise was epidemic in the 1990s when it became a buzz-modifier in various popularized social studies such as business school literature (e.g. Senge 1990; Argyris 1999; Easterby-Smith, Burgoyne, and Araujo 1999; 2000; Flood 1999; Garvin 2000). Senge was one of the earliest, although Argyris and Schön's exceptionally early riposte to Cyert and March's (1963) even earlier classic on 'single loop learning' is noteworthy. Argyris and Schön (1978; 1996) articulated their 'double-loop learning' approach in evolutionary systems language beyond the linear view of Cyert and March (1963) who described how organizations learn by changing practice short-term or long-term according to specific or general feedback rules respect-ively. This was very early 'systems thinking'. In Argyris and Schön's (1978) conception, single-loop feedback is seen as repetitively simplistic, while what they described as double-loop learning was based on organizational learning. An example of a thermostat is single-loop action, while changing thermostat performance involves learning. Noticeably many 'learning organization', 'learning company' and the like book and article titles were rushed into publication at or before the turn of the millennium.

This was significant sociologically because the social sensibility for change was 'in the air' as the leading economist of the previous millennium Alfred Marshall (1890) described the location of the 'mysteries of industry' residing, 'where children learn many of them unconsciously'. Accordingly, Peter Senge et al. (1999) one of the main gurus of 'learning organization' thinking later called his team's end-of-millennium book 'The Dance of Change'. Not only was the century changing, turbo-charged neo-liberal economics had arisen, just as laissez faire economics had done in Marshall's time, state-planned mixed or command economies had suddenly gone out of fashion with the end of the Cold War, the Internet heralded disruptive change across many societal, cultural, political and economic fields, though few were concretely visible yet, and global warming was beginning to be understood as potentially fatal to human society, at least in limited scientific circles. Resonating this sense of potentially unregulated, even uncontrollable change, governmental politics, as usual, 'blamed the victim' by issuing injunctions that society and, by inference its individuals, all had to learn how to accommodate to these existential changes or threats.

Social learning, in which this contribution is particularly interested, has four categories; the individual; the group; the organization; and society. Argyris and Schön's (1978) path-breaking organizational learning study still influences contemporary social learning psychology referencing contemporary practical applications (Von Schönfeld et al. 2020). That article notes how much social learning theory (SLT) presents as contributing largely to benign outcomes. It is thus consistent with psychology's therapeutic pretensions. As we shall elaborate below even the focus on how individuals and groups learn to behave in certain (more controlled) ways intends to influence practice towards psychologically balanced or rational outcomes from the normative perspective. Notice also how difficult it is to separate individual or group taxons from organizational or societal ones. This becomes extremely clear in our sub-section on the addictive effects of today's social media on 'mimetic learning' in each category. Thus a desired outcome enhancing, for instance, a socially inclusive result could equally have (intended or unintended) socially exclusive effects. Bandura (1977) departing from behaviourist towards cognitive learning theories confirmed the ability of humans to acquire new behaviour through observation and imitation by exclusive as well as inclusive actions. Argyris and Schön (1978), from an organizational studies perspective viewed social learning justified from a mitigation of business errors viewpoint advocating (reflexive) 'double-loop learning' to correct systemic mismanagement. But they too recognized Bandura's inference that the interests of the firm do not always coincide with the goals of wider society. The once widespread concern with 'sustainability' in advocating minimization of waste is nowadays criticized for advocating conservation of fossil fuels for future generations to exploit rather than ceasing their utilization full stop. To conclude this introduction having clarified the ensuing paper's line of interest, we realize that reference has been made explicitly to social, behavioural, cognitive and, implicitly, developmental learning theory. This is our contribution to the collection's quest for any meta-learning model.

Social learning theory – attention, addiction, aggression

A good example of a contribution with established provenance in both Cognitive and Developmental learning theory is Bandura's 'social learning theory of aggression'. We introduce this as it relates usefully to Girard's (1987) foreshadowed learning theory of Narcissism and its role in the dark psychology of attentional and addictive effects of contemporary psychopathologies of social media use. Bandura's by now dated analysis of aggression stresses the coercive activities of collective action agencies, notably: 'Agencies of government [that] are entrusted with considerable rewarding and coercive power' (Bandura 1983, 2–4). However, while aspects of that statement remain true, it is the 'rewarding and coercive' power of contemporary corporations and specifically those in private control of the 'surveillance' and 'attention' social media corporations that introduce a pernicious meaning to the 'learning' modifier. 'In the social learning view, people are endowed with neurophysiological mechanisms that enable them to behave aggressively, but the activation of these mechanisms depends on appropriate stimulation and is subject to cognitive control' (Bandura 1983, 5). The analysis begins with reference to the organism's capability of acquiring integrated behaviour patterns without constant updating through trial and error. Learning by observation consists in four inter-related sub-processes:

1. Attentional procedures regulate exploration and perception of observed activities transformed symbolically as images, words or signs.
2. Memory Representation performs guidance of symbolic images towards actions.
3. Motor Production procedures model these where SLT drives acquisition and activities of learned behaviours.
4. Incentive and Motivational procedures regulate performance of observationally learned responses.

In reference to contemporary Attention exploitation practices by data-miners the following is of interest. A key source of aggressive practice is the abundant symbolic modelling projected by the mass media. Research shows regular exposure to televised violence may have four effects. First, it facilitates learning about aggressive conduct. Second, it weakens restraints over aggressive behaviour. Third, it de-sensitizes and habituates observers to violent acts. Fourth, it shapes and models images of reality on which imitative actions are based (Bandura 1983, 7). This represents a post-behaviourist advocacy of SLT, known later also as social cognitive theory. As such, it opened up two psychological avenues that help understanding of the relative strength of 'attention' and 'imitation' as drivers of addictive performative practices (aggressive or obedient narcissism) learned by habitual observation of social media. By contrast, it also allowed revaluation of the benign rather than malign power of SLT in providing exemplary learning models.

First, however, we explore a certain fuzziness remaining at least superficially and possibly confusingly in the contrasting interest of SLT to be dyadic about the 'individual' and the 'social' as its focus (Edmondson 2011). To see learning as a response to individual failure (rather than success) is pretty straightforward and has been welcomed as a new cognitive model for aspirant entrepreneurs. The trick is to get entrepreneurial or any other 'victim' of failure to reflect on their *malpractice* with injunctions to avoid similar mistakes in the future. This may involve a review team to report on what happened and then distribute it throughout the business or other significant group or organization. In sport – especially – 'learning from failure' is eulogized as the Michael Jordan platitude after admitting to failing with 9000 shots at the hoop: 'Failure is good because you learn from it and you come back better'. Hitting 1000 tennis balls a day is a truism for minimal demonstration of aspirant competence for competitiveness in that championship sport (Larsen and Guggenheimer 2013). The criticism, though, is that the avatars who achieved success were also exceptional 'outlier' talents. Such 'celebrities' perform the role of 'clickbait' in the 'attention-to-addiction' spectrum of social media learning theory. At the individual level and for those non-celebrities terrified of being perceived as 'nonentities' the opponent is more likely to have a winning record against which the contender has to prevail. Even if 'double-loop' learning (changing the thermostat setting) from past failure occurs, psychological or physical limitations may contribute more than constant practice, or at the cognitive level, revision of strategy including 'imitation'. Psychological research advises that success enhances self-esteem, ego and vanity whereas failure represents humiliation (Jiang and Ngien 2020; Leary and Downs 1995). We shall see how the equivalent injunctions for groups and even whole societies to emulate 'successful' societies, economies or administrations by 'learning from failure' to be as successful as comparators, are similarly afflicted with a form of 'policy clickbait' syndrome (e.g. 'levelling-up').

Edmondson (2011) argues that from a business viewpoint these platitudes are misguided for three reasons. First, in organizational practice failure can be inevitable (as with the less-talented sportsperson) where novel 'shocks' occur, or complexity may create unpredictability that defeats resilience precautions from which new 'learning' is necessitated. Second, some failures can be considered bad if accepted and hitherto effective procedures were not adhered to or otherwise deviated from. However, such process deviations can be remedied or anticipated as in 'predictive management' (Cooke 2021). Thus in Toyota the andon cord can be activated by the line-worker where diagnosis and repair is allowed one minute for resolution or the line halted if not. Third, failure may be good where they are 'intelligent failures' that facilitate innovative solutions to novel problems. Even where robust testing regimes exist, as in pharmaceuticals or aerospace, never-before experienced resilience shocks may occur occasioning experimentation for new learning to restore predictability until the next time.

Consider how much more complexity is involved where 'policy clickbait' rests on injunctions that 'lessons must be learnt' in order to improve and even supersede some quality desired by policy leadership teams. School examination performance has been one public field where annual international benchmarking creates such pressures. Performance of health care systems – notably – during the Covid-19 episode – included various 'blame games' from government ministers aimed at local ambulance drivers, reaching to geographical comparisons of regional or federal domestic performance as well as international achievements prominent. This has long been sporadically if less systemically practised for the comparative geographic performance of economic regions and countries. The 'omnicrises' occurring in many countries from policing to energy supply and food and other supply chains have created significantly higher pressure on the 'learning of lessons' as the aftermath of causal events such as Covid-19, ranging to global semiconductor shortages, to Brexit. Many if not all these arise from the inadequacies of systems functioning with inadequate 'predictive management'. This is a protean form of Artificial Intelligence (AI) inspired approach to facilitating predictability into complex systems by integrating them into a platform operated by 'machine learning' (Cooke 2021). Here, algorithms effect learning by mimicking human and post-human puzzle-solving mechanisms (neural networks).

Learning: from social via economic to regional

Moving on to the less mechanistic 'mimesis' of individualistic cognition to the substantively more complex levels of SLT in 'collective action', the 'learning modifier' became of interest in the second half of the 1990s and 2000s. Both sought ambitiously to re-track the developmental trajectories and contexts (groups, classes, societies) by exploring the possibilities of the 'learning society' (Hughes and Tight 1995) along the economic geography parallel line studying 'learning regions' (Florida, 1995; Asheim, 1996; Morgan, 1997; Rutten, 2007, Simmie, 2011). Disappointingly the promise of this notion never materialized, partly because as Simmie's review shows it got bifurcated into a normative idea resting on the injunction that learning was a desirable end for 'regions' to aspire to, on the one hand, and more empirically that it was a modest action line in regional innovation strategy, on the other. Accordingly, it never developed analytically even though, as much of the regional innovation systems literature makes clear, there is evidently an acutely perceived need for 'policy crowdsourcing' for better 'levelling-up' of economic prosperity. Aspects of this highlighted better qualities of 'interactive learning' by firms and other innovation actors and for 'regional policy learning' to tackle issues such as 'cognitive dissonance' among corporate functionaries and entrepreneurs; 'convention analysis' of regional production culture; policy mixing to stimulate 'path creation'; and the hybrid skills to facilitate relatedness and transversality among policy functionaries (Cooke, 2012). Correctly, in its origins, in the work of Florida in 1995, 'learning region' is seen as a response to the rise of the knowledge economy as echoed by Asheim (1996) which grounds the notion in Lundvall & Johnson's (1994) advocacy of building a 'learning economy' to face the predicted tensions induced by the 'knowledge economy' paradigm. Finally, an application of the 'learning region' also occurred in Morgan (1997). This stressed the key to regional regeneration and improved social welfare lay in strengthening a region's social capital and institutional capacity to support learning.

Critique of the concept ranged from ascribing it the conceptual status of 'fuzzy', an 'impressionistic neologism', 'unlikely', 'over-localized' and challenged by 'learning asymmetries' (see, for example, Martin 2001). This is reminiscent of the many critiques aimed at the cognate concept of 'organizational learning' in the large corporation. Here problems concerning how to sample from where or whom to learn, whether what might be learned was applicable, and indeed whether it was yesterday's knowledge, meaning the learner would be engaged in a permanent failure to 'catch up' were all raised. This was the thinking in revisiting the concept by Rutten (2007). However, the kind of conceptual and policy instruments needed to achieve such endogenous regional change remained to be clearly specified – except in the desiccated modus operandi of 'machine learning'. We shall see below how some progress in this regard has been made from the viewpoint of complexity theory (Mitleton-Kelly 2006; 2011). Here, a key problem with the learning adjective is that it is linear, locking the learner into tuition of yesterday's path dependence rather than evolving in a non-linear fashion, seeking better outcome potential in what are known in complexity theory as 'strange attractors'. Such coalescences are often at the heart of innovations rather than just 'learnings'. She (Mitleton-Kelly 2006) understands why the 'learning' problem-framing for organizational re-design often produces disappointing results is due to the systemic nature of the task set, which is not really reducible, thus 'best practice' cannot truly be 'imitated' for re-assemblage:

Her take on complex co-evolving systems was their capacity for creating new order, notably new ways of working or relating. This further clarified the theoretical problem as a transition from thinking of path dependence to one of conceptualizing path interdependence, otherwise innovating through investigating the nature of 'strange attractors'. In comparative research on hospital change management, this approach was deemed superior to 'business process re-engineering' (Vakola & Rezgui, 2000) which reduced re-design to merely achieving financial equilibrium by so-called 'value-engineering' (or cost-reduction) which reduced re-design to merely achieving financial equilibrium. However, in advocating that as the prime aspiration the revised system was reductively driven to 'the edge of chaos' due to the reduction of 'organizational slack'. Reversing that by pilot-testing representative, emergent, co-evolutionary, socially nuanced and tolerant practices to bridge 'structural holes' (Burt 1992) and identify even 'strange attractors', a better solution ensued.

One reason why 'learning region' (Print Media Indicator of 60 at peak; 25 today), 'learning society' (PMI of 320 at peak; 85 today) and, to a lesser extent 'knowledge economy' (PMI of 10,250 at peak; 7,000 today; Örtenblad, 2022) entered narrative decline, registering a slide from the 1990s peak to 2022's current situation, is clear. It is because, as Ozga and Lingard (2006) perceived it, they were part of a global discourse promoted by OECD, the EU and the World Bank *inter alia* connecting each to economic competitiveness thus monetising the right to education (Stiglitz, 1999; Mulderrig, 2007; Patrick, 2013). Hence the linkage between the rise of a 'knowledge economy' and the need for an actively 'learning society' to fuel it expressed neoliberal politico-economic strategies on a global scale. Among criticisms of SLT specifically in a social media era are the following:

(1) Loss of knowledge control. Social learning implies teaching and learning. How 'expert' are the interactants?
(2) Limited data to show that social learning works.
(3) Learning on social media hinders productivity. Social media wastes time.
(4) Limited foreknowledge limits the scope of learners.
(5) 'Leadership' variability defocuses discussion.
(6) SLT learning is difficult absent: attention, retention, reproduction and motivation.
(7) Privacy and Security risks (Edmondson 2011).

These caused learning of the kind under discussion in this contribution to have been occluded by these and other concerns about learning under social media conditions. Item (6) above features highly among today's critiques of online learning where lack of self-motivation is particularly problematic. Absence of interactive face time with instructors scores high on learner dissatisfaction. Online learning is not adaptable compared to interactive learning. Online learning is seldom technically superior to live lecture contact. Online learning is likely to be unavailable for some key practical activities.

Despite such caveats, the ideology of 'smart' responses to contemporary arrests to developmental progress prevails. This includes digital re-design of functions through fashioning 'smart factories', 'smart offices', 'smart cities' and for the EU aspiring to allocate 'levelling-up'(populism UK-style) funding for regional budgets through 'smart specialization'. Each represents increasing reliance on algorithmic control and budgeting mechanisms that have the function of dehumanizing collaborative action. Smart factories (and kitchens) have been alluded to and the 'smart office' is its cognate 'dark facility' in the service sector where legal, administrative and even journalistic tasks can be fulfilled. 'Smart cities' are similarly designed for friction-free surveillance of consumer ID data-mining. This transforms their preferences in the form of persuasive, pervasive, attention-grabbing, digitized personal advertisements. Infrastructures for these architectural 'loot-boxes' are fashioned by designers like Norman Foster & Partners for Apple's digital campus or Heatherwick Studio for Google's parent firm Alphabet's affiliate Sidewalk Labs, notoriously in Toronto's – failed – Quayside 'smart neighbourhood'.

'Learning' with benign intentions but little experience of digital era manipulation

When the likes of Lundvall and Johnson (1994) fashioned their elaboration on anticipated anxieties raised by the 'knowledge economy' into a model response recommending the crafting of a 'learning economy', they did so with wholly benign intent.

> The approach emphasizes interdependence and non-linearity. This is based on the understanding that firms normally do not innovate in isolation but interact more or less closely with other organizations, through complex relations that are often characterized by reciprocity and feedback mechanisms in several loops. (Lundvall & Johnson, 1994)

Little could they have realized that a different, free market, neoliberal and individualistic model of 'market competition' would come to prevail by the turn of the century. Perhaps they should have, because key books heralding this had long-existed. The first, published in 1972, included a chapter by Nobel-Rijksbank economics laureate Herbert Simon which prefigured the 'Attention Economy':

> … in an information-rich world, the wealth of information means a dearth of something else: a scarcity of whatever it is that information consumes. What information consumes is rather obvious: it consumes the attention of its recipients. (emphasis added). (Simon 1972)

By the next century, once the aforementioned data-miners had absorbed Bandura's (1983) 'attentional' STL model it became fundamental. Traditional media advertisers followed a related model that suggests consumers went through a linear process they called 'Attention', 'Interest', 'Desire' and 'Action' (AIDA; Mithun and Hawaldar 2018). Attention is therefore the key stage in the process of converting potential customers. As costs of advertising became vanishingly low per capita due to online advertising, the consumer's 'attention' became Simon's 'dearth' or scarce resource to be allocated rather than 'information' as such.

The second book emphasizes egotistic psychology as a motivator of 'learning' with debateable, even 'psychopathic' social media outcomes, as we shall see. The author is Stanford University's René Girard (1987) and the subject is Girard's theory of mimetic desire – of people learning to want the same thing – that attempts to explain the origins of social conflict and violence. This clearly differs from Bandura's (1983) SLT of aggression, which is external and institutional whereas Girard's is internal, individualistic but having interactive effects on digital 'others'. Girard's fundamental concept of 'mimetic desire' highlights humanity as the species most apt at imitation. Indeed, he contends that imitation is the basic mechanism of learning. We learn inasmuch as we imitate what our teachers do, something neuroscience reports humans are suitably adapted to through the functioning of 'mirror neurons'. However, most research devoted to imitation paid little attention to the fact humans also imitate other people's desires, and depending on how this happens, it may lead to conflicts and rivalries. So, where Bandura's source of aggression is oppressive in origin, Girard's is replicative. Research discussed already (Jiang and Ngien 2020; Leary and Downs 1995) echoes Littrell's (2020) study that distinguishes 'grandiose' from 'vulnerable' narcissism. Thus success enhances self-esteem but not critical reflexivity whereas failure leaves subjects more insecure, defensive and introverted with lower self-esteem. Accordingly, grandiose narcissists feel more entitled while vulnerable narcissists feel more fearful with fragile self-esteem. Woods and Scott (2016) found night-time social media activity by teenagers reinforced low self-esteem while Vogel et al. (2014) discovered similar results for Facebook users. To conclude with the implications relating to Girard's exploration of learning mimetic desire it is likely that either category could be inclined towards 'mimesis' which has aggressive implications but that 'imitators' are less prone to harming others or self-harm, while 'mimetics' may display negative aspects of rivalry, including frustration, possible hostility and harmful conflict.

Aggressive, hostile trolling on social media is thought to have risen greatly as social media practice has been learned to a level of frequent aggressiveness by individual users and producers of such content under 'attentional' filtering. Vulnerable narcissism conditions are fomented by social media network algorithms. Of high political concern among societal leaders are trolling anonymity, aggressive content and woeful network policing on the part of providers. It is likely that attacks, knifings and murders of political representatives having occurred, regulatory controls on social media liberties will be tightened considerably. Even milder forms of learning of the kind to be described next are widely considered to be harmful to younger person self-esteem but tend to go with the grain of neoliberal capitalism, despite their tawdriness and are less immediate though still strongly psychopathically inclined towards vulnerable social users. Such practices amplify the gross inequalities arising from the role of 'influencers', especially 'bad influencers'. These perform another insidiously 'learned' practice of the 'attention economy' and its glorification of 'celebrity', 'branding' and 'vanity' (defined as futility or worthless pride compared to 'narcissism' which is a form of psychological ego disorder). Figure 1 is a representation of the 'Cognitive Supply Chain' of elements that drive 'celebrification' or by contrast 'humiliation'. We read from the top, noting two key drivers of the process starting with 'Attention' after Franck (2016) and Surveillance after Zuboff (2019).

On the left interaction link in the chain, 'Vanity' is highlighted as taught at Stanford's other 'influencer' institute, BJ Fogg's Persuasive Technology Lab where Instagram,

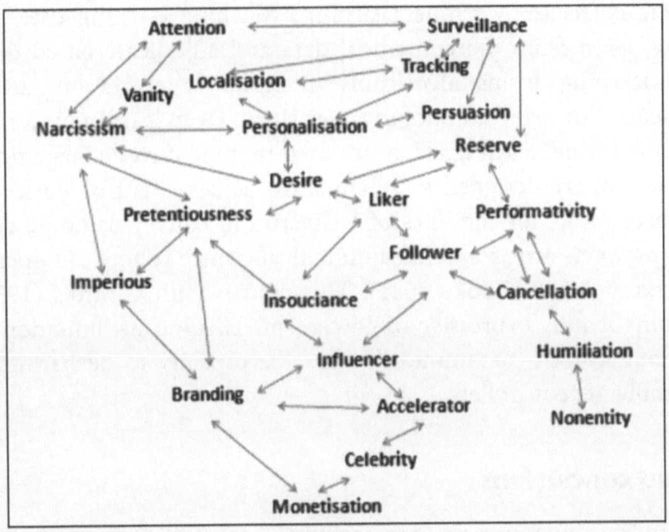

Figure 1. Pattern recognition diagram of a learned cognitive supply chain model.

WhatsApp and Twitter alumni were trained in Narcissism. As, then, libertarians before departing their Facebook and Google acquisition fates, such founders were 'dark entrepreneurs' insouciant of attitude and imperious of concern. Their online presence targeted 'Hard Branding' of products to attract 'Influencers' whose 'Celebrity' monetized their product sponsorship. For those 'Influencers' failing to reach 'Celebrity' status they remained 'Likers', 'Followers' or possibly sufferers of 'Humiliation' which might foment 'Vulnerability' or 'mimetic' trolling and worse. Among the many risks associated with such 'digital learning' apart from those mentioned, are the dark side of 'Influencer' status which turns into 'Bad Influencer'. Tim Wu (2017) described Kim Karshadian's 'pretentious' failure to 'Break the Internet' with her 55 million Instagram 'followers'. At least one of these, relatively unknown Jen Selter, then imitated her body sculpture, which in turn yielded 8.5 million Instagram followers. That led to sponsorship, endorsements and appearances as a newly successful 'attention merchant'. Later she was 'disembarked' from a flight after abusing a steward, claiming 'mental health' issues for a 'panic attack'. A more notorious case was that of Belle Gibson the Australian 'wellness influencer' for the 'clean food' diet and homeopathic remedies she claimed helped her 'beat cancer'. When her 'The Whole Pantry' blog was published it immediately garnered 200,000 'Likes' leading to also being posted on Apple Watch's 2014 launch, followed by a Penguin Books contract; cookbook published in 2014. However sceptics, notably Richard Guilliatt, were unconvinced and further pressure led Belle to recant all her 'cancer' claims. Her actions were described as 'particularly predatory and deceit on a grand scale for private profit' in the Australian courts. The due diligence competence of Apple and Penguin were later also scrutinized and critiqued but revealed little or no contrition or promise that 'lessons would be learned'.

So finally, what is the fate of 'Learning' today? The clear answer is that under the twin influences of 'Surveillance Capitalism' (Zuboff, 2019) in the 'Attention Economy' (Wu 2017) artificial intelligence (AI) and robotics have created 'Learning Machines' and their algorithmic correlate 'Machine Learning'. Machine learning uses algorithms to summarize data, learn from the content of data and adjudicate based on what it has learned. Deep learning forms algorithms in strata to create an 'artificial neural network' that 'learns' to make autonomous decisions. Deep learning is a more sophisticated version of machine learning. They are used in 'predictive management' fashioned by 'digital twins' that are designed to facilitate the achievement of 'dark' or 'unlit' factories. These serve meals for the likes of Deliveroo to convey to home consumers. In 'dark' factories products are assembled automatically since robots can operate normally without industrial lighting (Cooke 2021). This chimes with Girard's (1987) view that 'imitative' learning is fully expressive of narcissism. This means 'imitation' is incapable of innovation, but worse – as 'mimesis' – has the capacity to be harmful both to its users and, arguably, its controllers.

Discussion and conclusions

The 'dark' side of learning represented in 'trolling', 'hate mail' and online abuse has led to re-appraisal of its practical desirability in a digital era. One of the noteworthy features of the enveloping neoliberal ontology accompanying the rise of social media, AI, robotics and 'smart' ubiquity is the phobia against foresight, testing or planning 'solutions'. 'Solutionism' as Bridle (2018) calls it, is berated as 'computational thinking' – reductionist, linear and 'solutionist' – source of the system ignorance most modern humans experience and display. Hence, the corollary of distrust of expertise, foresight, testing or planning has come to be true. Accordingly, where 'solutions' can no longer be 'planned', or 'learned' about, the problem can apparently be ignored, inaction therefore can prevail or at best be delayed until crisis or catastrophe reveals that 'the market' has failed. This is Bridle's somewhat dystopian 'New Dark Age' in which the point is that 'learning' became redundant, if not obsolescent. Market 'spontaneity' as Hayek referred to it (Tao 2016), creates its own reality. The market, said Hayek, was a spontaneous order. By spontaneous he meant unplanned – the market was not designed by anyone but evolved slowly as the result of human actions. Recent thought on this has extended beyond the egotistic individual to the fashionable neoliberal conceit that 'ecosystems' are 'spontaneous orders' which seems to allow 'learning' in through the back door, especially in reference to the field of environmental ethics (Lamey 2015).

Neoliberal 'spontaneity' was embedded in the knee-jerk reactions of the UK finance minister George Osborne, whose policy actions were described by the press as an 'omnishambles' failing to introduce pernicious 'austerity' measures to UK society following the financial crash of 2007–2008. From the Trump administration's blanket response of denial coated with otiose hyperbole ('200,000 would be a good number') faced with US Covid-19 deaths rising by October 2021 to over 725,835, to UK leader Boris Johnson's claims that the UK's Covid-19 'test and trace' system was 'world-beating' when it simply malfunctioned, rational thought was widely jettisoned. In the UK, constant and humiliating 'U-turns' were announced by ministers based on hope informed by hunch and regular wrongful claims in fields as diverse as: prime ministerial adviser Dominic Cummings, a historian, mis-comparing government to the human immune system, failing to see it as adaptive but not as also structural; Johnson's uncritical reification of 'science' by which his government was claimed to be led, supinely, as it transpired; UK complacency and boosterism in fatally delaying early lockdown in Spring 2020; assuming the pandemic was flu-like not sui generis; protecting the National Health Service (NHS) from hospitals being overwhelmed but forgetting critically ill non Covid-19 patients and care homes where thousands died of the virus; mis-deployment of standardizing algorithms giving inaccurate examination results to all sitting candidates; in all these and many other cases, recantation of mistaken policy announcements was forced on government by public and press outrage (McTague 2020).

So, to conclude, we may make three main conclusions. The first is that 'learning' became a widely applied modifier caused by the rise of a perceived 'knowledge economy' in the 1980s. This created anxiety at the political level causing much concern that 'lifelong learning' was needed to confront economic change. Second, it coincided with a counter-reaction caused implicitly by neoliberal politics eschewing 'planning' or, apparently 'learning' after the end of the Cold War under the influence of Hayekian monetarist ideology. Both of these main forces are being slowly undermined as the realization of perceived benign interpretations of the digital economy and 'spontaneous' markets as the only source and apotheosis of rational change, have been shown to have deleterious, anti-social and dangerously pernicious effects on society's physical and psychological health. The case for a revived and re-regulated 'learning' arena in modern jurisdictional, administrative and economic affairs is thus overwhelming.

Disclosure statement

No potential conflict of interest was reported by the author(s).

References

Argyris, C. 1999. *On Organizational Learning*. 2nd ed. Oxford: Blackwell.

Argyris, C., and D. Schön. 1978. *Organizational Learning: A Theory of Action Perspective*. Reading, MA: Addison Wesley.

Asheim, B. 1996. "Industrial Districts as Learning Regions: A Condition for Prosperity." *European Planning Studies* 4 (4): 379–400.

Argyris, C., and D. Schön. 1996. *Organizational Learning II: Theory, Method and Practice*. Reading, MA: Addison Wesley.

Bandura, A. 1977. *Social Learning Theory*. New York: Prentice-Hall.

Bandura, A. 1983. "Psychological Mechanisms of Aggression." In *Aggression, Theoretical and Empirical Reviews*, edited by R. Geen and E. Donnerstein. New York: The Academic Press.

Bridle, J. 2018. *New Dark Age: Technology and the End of the Future*. London: Verso.

Burt, R. 1992. *Structural Holes: The Social Structure of Competition*. Cambridge, MA: Harvard University Press.

Cooke, P. (2012) *Complex Adaptive Innovation Systems*. London: Routledge.

Cooke, P. 2021. "Image and Reality: 'Digital Twins' in Smart Factory Automotive Process Innovation – Critical Issues." *Regional Studies* 55 (10–11): 1630–1641. doi:10.1080/00343404.2021.1959544.

Cyert, R., and J. March. 1963. *A Behavioral Theory of the Firm*. NJ: Prentice-Hall.

Easterby-Smith, M., J. Burgoyne, and L. Araujo. 1999. *Organizational Learning and the Learning Organization*. London: Sage.

Easterby-Smith, M., M. Crossan, and D. Nicolini. 2000. "Organizational Learning: Debates Past, Present and Future." *Journal of Management Studies* 37 (6): 783–796.

Edmondson, A. 2011. "Strategies for Learning from Failure, Harvard Business Review", Reprint R1104B, April.
Flood, R. 1999. *Rethinking the Fifth Discipline: Learning Within the Unknowable*. London: Routledge.
Florida, R. 1995. Toward the Learning Region. *Futures* 27 (5): 527–36.
Franck, G. 2016. "Vanity Fairs: Competition in the Service of Self-esteem." *Mind & Matter* 14 (2): 155–165.
Garvin, D. 2000. *Learning in Action: A Guide to Putting the Learning Organization to Work*. Boston, MA: Harvard Business School Press.
Girard, R. 1987. *Things Hidden Since the Foundation of the World*. Stanford: Stanford University Press.
Hughes, C., and M. Tight. 1995. "The Myth of the Learning Society." *British Journal of Education Studies* 43 (3), doi:10.1080/00071005.1995.9974038.
Jiang, S., and A. Ngien. 2020. "The Effects of Instagram Use, Social Comparison, and Self-Esteem on Social Anxiety: A Survey Study in Singapore." *Social Media & Society* 6 (2). doi:10.1177/2056305120912488.
Lamey, A. 2015. "Ecosystems as Spontaneous Orders." *Critical Review: A Journal of Politics and Society* 27 (1): 64–88. doi:10.1080/08913811.2014.947738.
Larsen, E., and J. Guggenheimer. 2013. "The Effects of Scaling Tennis Equipment on the Forehand Groundstroke Performance of Children." *Journal of Sports Science & Medicine* 12 (2): 323–331.
Leary, M., and D. Downs. 1995. "Interpersonal Functions of the Self-esteem Motive." In *Efficacy, Agency, and Self-Esteem*, edited by M. Kernis. Boston, MA: Springer. doi:10.1007/978-1-4899-1280-0_7.
Littrell, S., J. Fugelsang, and E. Risko. 2020. "Overconfidently Underthinking: Narcissism Negatively Predicts Cognitive Reflection." *Thinking and Reasoning* 36 (3): 352–80.
Lundvall, B., and B. Johnson. 1994. "The Learning Economy." *Journal of Industry Studies* 1 (2): 23–42.
Marshall, A. 1890. *Principles of Economics*. London: Macmillan.
Martin, R. 2001. "Geography and Public Policy: The Case of the Missing Agenda." *Progress in Human Geography* 25 (2): 189–210. doi:10.1191/030913201678580476.
McTague, T. 2020. "Covid Revealed Sickness at the Heart of Britain", The Times, August 15, 30–31.
Mithun, U., and I. Hawaldar. 2018. "Influence of Advertisements on Customers Based on AIDA Model." *Problems and Perspectives in Management* 16 (4): 285–298. doi:10.21511/ppm.16(4).2018.24.
Mitleton-Kelly, E. 2006. "A Complexity Approach to co-Creating an Innovative Environment." *World Futures: The Journal of General Evolution, Special Issue* 62 (3): 223–239.
Mitleton-Kelly, E. 2011. "A Complexity Theory Approach to Sustainability: A Longitudinal Study in two London NHS Hospitals." *The Learning Organization* 18 (1): 45–53.
Morgan, K. 1997. "The Learning Region: Institutions, Innovation and Regional Renewal." *Regional Studies* 31 (5): 491–503.
Mulderrig, J. 2007. "Textual Strategies of Representation and Legitimation in New Labour Policy Discourse." In *Marxism and Education: Renewing Dialogues*, Vol. 1, edited by G. Rikowski, A. Green, and H. Raduntz, 135–150. New York: Palgrave Macmillan.
Ozga, J., and R. Lingard. 2006. "Globalisation, Education Policy and Politics." In *The Routledge Reader in Education Policy and Politics*, edited by R. Lingard and J. Ozga. London: Routledge.
Patrick, F. 2013. "Neoliberalism, the Knowledge Economy, and the Learner: Challenging the Inevitability of the Commodified Self as an Outcome of Education." *International Scholarly Research Notices* 2013, doi:10.1155/2013/108705.
Rutten, R. 2007. *The Learning Region: Foundations, State of the Art, Future*. Cheltenham: Edward Elgar.
Senge, P. 1990. *The Fifth Discipline; The Art and Practice of the Learning Organization*. London: Century Business.
Senge, P., A. Kleiner, R. Ross, G. Roth, and B. Smith. 1999. *The Dance of Change*. New York: Currency Doubleday.

Simmie, J. 2011. Learning Regions. In *Handbook of Regional Innovation and Growth*, edited by P. Cooke, B. Asheim, R. Boschma, R. Martin, D. Schwartz, and F. Tödtling, 547-55. Cheltenham: Edward Elgar Publishing.

Simon, H. 1972. "Designing Organizations for an Information-rich World." In *Computers, Communication, and the Public Interest*, edited by M. Greenberger, 37–52. Baltimore, MD: Johns Hopkins Press.

Stiglitz, J. 1999. *Public Policy for a Knowledge Economy. Speech at Centre for Economic Policy Research*. London.

Tao, Y. 2016. "Spontaneous Economic Order." *Journal of Evolutionary Economics* 26 (3): 467–500.

Vogel, E.., J.. Rose, L.. Roberts, and K. Eckles. 2014. "Social Comparison, Social Media, and Self-esteem." *Psychology of Popular Media Culture* 3 (4): 206. doi:10.1037/ppm0000047.

Vakola, M., and Y. Rezgui. 2000. Critique of existing business process re-engineering methodologies. *Business Process Management Journal* 6 (3): 238–250.

Von Schönfeld, K., W. Tan, C. Wiekens, and L. Janssen-Jansen. 2020. "Unpacking Social Learning in Planning: Who Learns What from Whom?" *Urban Research & Practice* 13 (4): 411–433. doi:10.1080/17535069.2019.1576216.

Woods, H., and H. Scott. 2016. "Sleepyteens: Social Media Use in Adolescence is Associated with Poor Sleep Quality, Anxiety, Depression and Low Self-Esteem." *Journal of Adolescence* 51 (1): 41–49. doi:10.1016/j.adolescence.2016.05.008.

Wu, T. 2017. *The Attention Merchants*. London: Atlantic.

Zuboff, S. 2019. *The Age of Surveillance Capitalism*. London: Profile.

Does regional innovation policy really work for Industry 4.0? Evidence for industrial districts

Jose-Luis Hervas-Oliver, Sofia Estelles-Miguel, Marta Peris-Ortiz and Jose Antonio Belso-Martínez

ABSTRACT

Industry 4.0 threatens established lock-in paradigms in some districts. In this study, we compare different innovation policies aimed at facilitating Industry 4.0 awareness and its adoption in three Marshallian Industrial Districts (MIDs), asking: 'What' types of innovation policies work for Industry 4.0 in industrial districts? And, where they do, 'how' are those innovation initiatives designed, developed and implemented for digitizing districts? Using qualitative evidence based on 24 interviews and the review of existing literature concerning Industry 4.0 in three MIDs, results show different mechanisms and approaches for creating awareness and maximizing the diffusion of Industry 4.0 in each district, as a result of each local 'cognitive structure'. One size-fits-all policies are not realistic for digitization: collective actors leading place-based collective actions that are bottom-up and co-designed with public and private local actors is what works best. For policymakers, this study presents guidance for developing Industry 4.0 in MID settings.

1. Introduction

Oesterreich and Teuteberg (2016) established that the term Industry 4.0 is comprised of a variety of enabling technologies (e.g. Cloud Computing, Internet of Things, Big Data, Artificial Intelligence, etc.) which seek to develop a digital and automated manufacturing environment as well as the digitization[1] of the value chain. Industry 4.0 in manufacturing and other business activities, alters companies, industries and value chains, as well as their associated production, distribution and consumption systems, provoking a digital disruption that has configured a new manufacturing paradigm. In this study, Industry 4.0 is the digitization of manufacturing due to the introduction of enabling technologies, such as Cyber-Physical systems, Big Data and others. Industry 4.0 is said to bring positive outcomes, such as new business models, better productivity and better sustainability, (e.g. Porter and Heppelmann 2014; Lepore and Spigarelli 2020; Teixeira and Tavares-Lehmann 2022).

While SMEs face difficulties in adopting Industry 4.0, especially due to their lack of resources (e.g. finance, IT systems and other resources; Matt et al. 2021; Yüksel 2020), SMEs in districts also present an additional obstacle: strong local networks and lock-in institutions (rules, norms, assumptions, paradigms, etc.) that prevent change and promote uniformity, along with avoiding radical changes (see Glasmeier 1991). The introduction of Industry 4.0 in regions and districts addresses a different system level of analysis that facilitates a more sensitive 'place-based' approach. Industry 4.0 in regions and districts remains under-researched (e.g. Lazzeretti et al. 2022; Pagano et al. 2021; Bettiol et al. 2021; Bellandi, de Propris, and Santini 2019; Bellandi et al. 2020; De Propris and Bailey 2020; Hervas-Oliver 2021), especially in terms of innovation policies. The latter constitutes the goal of this article. In this study we integrate both perspectives, analysing collective initiatives to introduce Industry 4.0 in districts and their local SMEs, approaching the micro- and meso-level of analysis.

In particular, this study researches the following questions: 'What' types of innovation policies work for Industry 4.0 in industrial districts? And, where they do, 'how' are those innovation initiatives designed, developed and implemented for introducing collectively Industry 4.0? This study analyses and compares Industry 4.0 initiatives to collectively introducing Industry 4.0 in industrial districts (MIDs).

This study's focus is the design and development of Industry 4.0 in three different districts in Spain: the 'Vinalopo' Footwear district, the Toys-Plastics district and the Ceramic tile district of Castellon. We perform a critical review and analysis of existent evidence in those districts complemented with additional empirical factors obtained by interviews.

In our study, the focal process is the analysis and comparison of different policy initiatives to foster Industry 4.0 adoption in MIDs, and the setting are three MIDs in Spain: the 'Vinalopo' Footwear district, the Toys-Plastics district and the Ceramic tile district of Castellon. Our access to the Industry 4.0 policy development process and its main actors, makes these districts well suited to our purpose. Following Eisenhardt and Graebner (2007), we develop case studies for theory-building, along with a critical review of literature on those districts, answering research questions that address 'how' and 'why' in unexplored research areas. Our study and method are both justified by the fact that although we observe that innovation policies in districts are ubiquitous, yet literature is poor on policies for Industry 4.0.

Our results contribute to understand the factors that support effective policies to support collectively the introduction of Industry 4.0 in MIDs (e.g. Bellandi, de Propris, and Santini 2019; Hervas-Oliver et al. 2019; Bellandi et al. 2020), contributing to industrial districts (e.g. Belussi and Sedita 2009; Hervas-Oliver, Belso-Martínez, and Díez-Vial 2022).

The remainder of the paper is organized as follows: After this introduction, Section 2 presents a literature review. Then, Section 3 shows methodology and data. Section 4 shows the district analysed and the empirical results comparing innovation policies. Finally, conclusions are presented in Section Five.

2. Literature review

2.1. Industry 4.0 in MIDs: what do we know?

2.1.1. SMEs in districts

MIDs are socio-economic contexts characterized by cooperation and competition among local SMEs in low-tech environments where 'innovation without research' is the usual business (Galletto and Domènech 2014; Hervas-Oliver et al., 2015; Boix, Galletto, and

Sforzi 2019), except for the large firms that orchestrate the local networks ('leading or technology gatekeepers' in the sense of Lazerson and Lorenzoni 1999). These leading firms diffuse knowledge within their networks of SMEs through intense network relationships based on trust, reputation, custom, reciprocity, reliability, and openness to learning that are pervasive in districts and districts (e.g. Munari, Sobrero, and Malipiero 2012). The local SMEs hardly perform any R&D activities, substituting them with intense cooperation, social ties and interactions with other local counterparts, including learning from and imitation of those leading firms. On the contrary, the leading firms are the ones performing R&D activities, as well as local support organizations, such as research institutes or universities. SMEs in this context might potentially suffer lock-in and cognitive inertia due to the excessive reliance on locally-based assumptions and paradigms, in no small part due to the fact that the local learning process is primarily based on 'incremental' innovation from knowledge interaction and exchange based on trust, social capital and repetitive interactions (e.g. Glasmeier 1991).

As regards 'Industry 4.0' of SMEs, as Moeuf et al. (2020) show, smaller companies present poorer integration of digital enablers and knowledge management in manufacturing facilities. As digitization is a capability in itself, related to IT and information systems, digitizing requires organization capabilities and knowledge management that enable learning and capabilities reconfiguration for introducing digital enablers in SME resources and capabilities. Such knowledge and the pre-conditioning of assets, especially those related to IT and software know-how that underpin digitization, however, are not particularly well developed in SMEs. Similarly, digitization also needs investments in machinery and facilities and supply chain re-organization, which are also more complicated for SMEs (Müller, Buliga, and Voigt 2018).

SMEs in districts, like other types of SMEs, do not know what to do about Industry 4.0 and other radical changes, due to their low level of capabilities to digitize and insufficient resources (Sommer 2015). As Matt et al. (2021) and Yüksel (2020) show, smaller companies present poorer integration of digital enablers and knowledge management in manufacturing facilities, that is, poor IT capabilities. Industry 4.0 is considered a disruption that involves radical organizational changes (see Bellandi et al. 2020), far beyond the local learning process based on incremental innovations. In the case of MIDs, SMEs are typically conducting incremental innovation lock-in technology and assumptions established by leading incumbents that organize local networks (see Munari, Sobrero, and Malipiero 2012). SMEs in districts are reluctant to change in front of radical innovations (see Glasmeier 1991; Hervás-Oliver et al. 2018; Hervas-Oliver 2021). Industry 4.0 is considered a disruption that involves radical organizational changes (see Bellandi et al. 2020), far beyond the local learning process based on incremental innovations. Corò and Volpe (2020) point out the difference between Industry 4.0 awareness and adoption, the latter at a very low rate (10% for Veneto SMEs). In addition, SMEs either in or outside districts, do not frequently engage with universities or research technology centres, preferring largely to cooperate and learn from other firms along their value chain (suppliers and customers).

Therefore, SMEs are very constrained in terms of adopting Industry 4.0, especially because the legitimization needed to be in the networks of learning orchestrated by leading firms. Radical changes are not easily welcomed in districts, where lock-in technologies, institutions and paradigms facilitate incremental, but discourage radical changes (Gilbert 2012; Hervás-Oliver et al. 2018). All these constraints, along with the usual ones when it comes to the digitization of SMEs (e.g. Moeuf et al. 2020; Müller, Buliga, and Voigt 2018) complicate the change towards digital technologies.

2.1.2. The district advantage for Industry 4.0: cooperation, collective actors and actions

SMEs in districts, nevertheless, have an additional advantage, due to the logic of competition–cooperation. Put differently, districts facilitate collective actions that can be very useful to signal change and show the way to digitize among local SMEs (see Hervas-Oliver 2021; Hervas-Oliver et al. 2019; Pagano et al. 2021). Bellandi, Chaminade, and Plechero points out this factor calling it the 'collective rerouting' in districts, facilitating the incorporation of new knowledge and the transformation of networks, all supported by existing social relations and an institutional local system. SMEs in districts, by developing collective actions, might transit towards digitizing using the power of local cooperation. This cooperation is effectively applied to real cases, such as 'open-factories' or 'living-labs' for co-creation of new digital technologies in districts that foster cooperation and knowledge recombination to attract SME imitation, even facilitating open-doors policies to spread the awareness of change and show the way to do it (see Hervas-Oliver et al. 2019; Pagano et al. 2021). The collective actions are led and orchestrated by local actors and are based on open innovation and collaborative co-design and implementation, sharing common visions of the local cognitive structure required to digitize.

In any case, Industry 4.0 in regions and districts is said to be a place-based phenomenon that requires adaptation to those local assumptions, technologies, local networks, cognitive structures and institutions. In this respect, Sandulli, Gimenez-Fernandez, and Rodriguez Ferradas (2021) empirically show how it is not possible to define a single 'one-size-fits-all' policy for promoting Industry 4.0 that is generalizable to all regions but that it is dependent on the specific characteristics of each regional innovation system and, therefore, each district. Even the European DIH initiative[2] from RIS3 is based on bottom-up initiatives where local actors co-organize and co-design their digitization activities.

2.2. Placed-based innovation policies

What is the nature of place-based innovation policies? We refer to place-based innovation policies (e.g. Barca, McCann, and Rodríguez-Pose 2012; Magro and Wilson 2019) as initiatives built upon locally-sensitive associative structures of governance that are bottom-up, decentralized, open and consultative. These initiatives also involve local actor coordination and allow the collaborative co-design of actions by public and private joint efforts aimed at achieving collective goals by sharing common understanding about the local territory and its potential change. Similarly, Feldman and Lowe (2018) posit that effective policies are based upon bottom-up endogenous negotiations among local actors that include subsequent adaptations and incremental changes in response to changing conditions. These initiatives are constructed upon creative actions and joint collective decisions that take into account local social conditions and the interactions of actors in the policy. These collective actions are based on social practice and interactive learning, building upon the basis of collectively shared understanding of a territory's strategic needs and priorities (Ebbekink and Lagendijk 2013, 749). These initiatives are oriented to solve future problems, present a diversity of multi-actors, and are oriented to joining complementary capabilities from different actors and industries, enticing cross-fertilization of ideas and promoting joint or collective action (e.g. Uotila, Harmaakorpi, and Hermans 2012). In a similar vein, Magro and Wilson (2019) refer to a similar concept, highlighting the open debate, trial and error and co-design of initiatives by different local stakeholders that cross-fertilize ideas to find a win-win solution in a collaborative atmosphere.

National initiatives for digitizing (e.g. 'HUBS, Catapults', etc.), as mentioned above, are based upon the idea of open innovation, collaboration and the development of networks for inter-firm interaction and innovation. For instance, the 'Digital Innovation Hubs' created by the European Union (see Hervas-Oliver et al. 2021) are one- stop-shops for SMEs, mid- caps and more mature or established companies, which provide a range of supporting services, including technology testing, financing advice, market intelligence, training, promotion and marketing, and networking opportunities for Industry 4.0 adoption. At the national level, digitization policies follow the principle of 'place-based initiatives', where location matters. For instance, the German 'Hubs' (defined as partnerships that connect SMEs, startups, corporation, research and science-based organizations for promoting digitization) located the leading 'Fintech' technology centre in Frankfurt, the financial capital, and the 'Mobility' platform in Munich and Bayern, the heart of the automotive industry in the country. Through these 'Hubs', the German government promotes alliances among universities, researchers, investors, institutions and companies, for creating digital innovation hubs sponsored by its national programme established in 2017 ('Digital Hub Initiative' from the 'Digitale Strategie 2025'). Similarly, the UK Innovate policy[3] fosters the digital transition through different programmes, such as the 'Catapult'[4] initiative that encourages the creation of connected centres in the country primarily aimed at converting the country's research and development into commercial products or solutions available for manufacturing firms. The programme is aimed at bridging the gap between business and academia, helping to turn great ideas into reality by providing access to world-class research and development facilities and expertise that would otherwise be out of reach for many businesses in the UK. In order to achieve this transformation, the initiative is based upon cooperation among companies, research workers and institutions (R&D centres, universities, accelerators, etc.), promoting the formation of networks and ecosystems to facilitate the application and commercialization of research.[5]

All in all, we posit that the design, development and implementation of collective actions based on cooperation in the digitization of districts becomes a very district-specific and place-based process. Thus, we point out that place-based initiatives for digitizing, need to embrace those above-mentioned principles such as being centred around place-based sensitive and common goals, open innovation, bottom-up co-design of initiatives, trial and error, involving local private and public actors in cooperation and the leadership of collective actors (see Hervás-Oliver 2021). By 'collective actor', we refer to public and private organizations formed by a coalition of industry, government and science representatives that are geographically, institutionally and socially embedded (e.g. York, Hargrave, and Pacheco 2016).

3. Method, data and settings

3.1. Methodology

This study utilizes a literature review (for the digitization of the Spanish Ceramic tile district, the Vinalopo Footwear and the Toy-Valley district), along with 24 additional interviews. A literature review concerning the introduction of Industry 4.0 in the Ceramic tile district (from Hervas-Oliver et al. 2019) is complemented by access to the empirical material (transcriptions, secondary reports, etc.; 30 interviews). Similarly, the approach to understanding innovation initiatives for digitization of the Toy Valley (from Hervás-Oliver 2021), is complemented by access to the transcriptions and material from the district (54 interviews during 2017 and 2020). In a similar way, for the Vinalopo Footwear district we follow suit, accessing the empirical material (45 interviews, from Hervás-Oliver 2022).

Lastly, in 2021, 24 additional interviews were carried out to evaluate innovation policies in each district. This qualitative approach through local business associations and RTIs leading digitization initiatives (8 in the Vinalopo Footwear, 8 in the Toy-Valley and 8 in the Ceramic tile district of Castellon), extend our knowledge on each district. Interviews include a basic current description of the districts, complemented by the Industry 4.0 policies implemented in each district. The three districts are considered traditional Marshallian Industrial Districts (see Boix and Galletto 2009, among many others).

3.2. Settings: industrial districts in the analysis

Each District's features are summarized and explained in Tables 1 and 2. See Tables 1 and 2.

As observed in both tables, the three districts analysed present common features of their MARK III status,[6] all of them showing a high degree of specialization, the significant presence of foreign (and indigenous) multinationals and a high-quality supportive structure of training and scientific research (e.g. local business associations, vocational training centres, research and technology transfer institutes devoted to local technologies and highly embedded in the territories, etc.). In particular, the interviews commented on the importance of the global value chains, the benefits of hosting foreign multinationals and the necessity to turn the tide towards digitization and sustainability.

These districts present vertical Research and Transfer Institutes (RTIs) that are well embedded in the territory and focused on local technologies. These RTIs ('ITC' at the Castellon ceramic tile; 'Inescop' at the Vinalopo footwear and, AIJU at the Toy-Valley) are world-class organizations in their fields that belong to the Valencian REDIT (network of institutes[7]) and enjoy coalitions of public and private stakeholders, being public organizations (in ownership) but privately run by a Board of Directors voted among the local firms in each district focused on. They present a mixed finance portfolio, with public (around 60%) and private (around 40%) funding, the latter form the services that are provided to firms (testing, product certification, R&D, etc.) and others (European Projects participation, other national competitive funding, etc.). These RTIs are 'public' labs that nurture districts with new technology and innovation for supporting the firms' innovation processes. They also provide support services

Table 1. Main districts' features I.

Districts	Description/type of district	Local supporting organizations and infrastructure	Local value chain and industries located
Footwear (Vinalopo, Alicante)	• Description: A vibrant world-class footwear production specialized in women shoes (high-heel and high-medium end). • Strengths: very good local supporting organizations, many multinationals operating, European hub for footwear-related business.	• RTI (Inescop) • Local business association (AVECAL) • Vocational training centres • Exhibitions, show rooms • Spanish Association of Footwear Components Firms (AEC) • Federation of Spanish Footwear Industries (FICE) • Others	Auxiliary industry: producers of footwear components, packaging for footwear, all different stages of footwear (design, R&D, production, logistics, marketing, etc.).
Ceramics (Castellon)	• Description: A vibrant world-class ceramic tile production and leading district for tile decoration (glazing firms in the chemistry industry for tiles); • Strengths: very good local supporting organizations, many multinationals operating, European hub for ceramic-tile related business.	• RTI (ITC) • Local business association (ceramic tile ASCER; cermic equipment, Asebec; ceramic decoration, Anffec) • Vocational training centres • Exhibitions, show rooms, fair trade (CEVISAMA) • Others	Auxiliary industry: producers of decoration (chemistry) and equipment for ceramic tiles, packaging for ceramics, all different stages of ceramic tile production (design, R&D, production, logistics, marketing, etc.).
Toys-Plastics (Ibi)	• Description: A vibrant world-class district for plastics-packaging and toys production • Strengths: specialized in moulding for plastic injection; very good local supporting organizations, many multinationals operating, European hub for packaging and plastics related business.	• RTI (AIJU) • Local business association (IBIAE) and cluster association (Toy Valley cluster association) • Vocational training centres • Exhibitions, show rooms • Spanish Association of Toy Producers (AEFJ) • Others	Auxiliary industry: producers of plastics, plastic components, packaging for different markets (pharmacy products, health care, beverages, automotive, furniture, toys and other industries), all different stages of plastic-packaging production (design, R&D, moulding, production, logistics, marketing, etc.).

Source: Own, based on literature.

Table 2. Main districts' features II.

District	Specialization Index **	Number of firms
Footwear (Vinalopo)	469% Exporting >60%	• Around 2,300 very small firms, predominantly around 1–9 employees • Around 30,000 manufacturing jobs in the local technology (footwear production and auxiliary industries)
Ceramics (Castellon)	338% Exporting >80%	• Around 120 tile producers plus 60 in auxiliary industries; small firms predominantly starting from 50–100 employees; • Around 20,000 manufacturing jobs in the local technologies
Toys-Plastics (Ibi)	170%+ Exporting >40%	• Around 400 firms, predominantly around 1–9 employees • Around 7,000 manufacturing jobs in the local technologies

Source: Own elaboration, also from interviews;** Valencian Region over Spanish (data in specific NACE activity codes), https://argos.gva.es/documents/165533218/172307874/Entregable+7_Especializaci%C3%B3n+productiva+de+la+Comunitat+Valenciana/e3e701d6-2bad-4753-a707-454a3aab57bf; for Toys is about 200%.

tailored to the districts' needs. The districts are also well represented by local business associations.

Each of the three districts displays a very different local value chain and division of labour, showing different concentration indexes (measured through 'Specialization

Indexes' using Spanish data according to specific NACE activity codes), firms and export intensities. See Tables 1 and 2 for details.

The local 'networks' display very interesting differences, showing different structures and mechanisms for knowledge diffusion. In the Castellon ceramic tile district, all firms manufacture the final product (ceramic tiles) and the equipment and chemical firms (production and decoration, respectively) are the ones transferring new knowledge and innovation. Despite the fact that all firms compete, we observe inter-firm cooperation not only in the producer-user spectrum (ceramic tile firms with equipment and chemical firms) but among ceramic tile firms that specialized in different types of products and then exchange them to complete product portfolios or to serve a single customer with a full range of products. Inter-firm cooperation among competitors, not only along the division of labour, is observed and also fosters an intense process of local imitation.

As one interviewee revealed at the ITC:

> The reality of this district is a very intense cooperation, not only in collective efforts but the way that informal groups based on trust and good relationship trade types of products among themselves, allowing for high specialization of firms. Having said that, competition is also very intense.

The Vinalopo footwear district is quite different, as not all firms produce the final product (footwear) but the different components (leather, soles, heels, fastening, rubber, etc.). The division of labour, therefore, is orchestrated by the leading firms that produce (coordinate the assembling of) the final product. These firms are also the ones that concentrate design, marketing and access to the distribution channels (even with their own physical or online stores). ZARA's subsidiary (Tempe) constitutes one example of these firms. Cooperation is non-intense among competitors. Cooperation, therefore, is especially observed along the division of labour (user-producer, outsourcing agreements) but in a quasi-hierarchical approach, as small producers of components are very dependent on those 'leading' firms.

The Inescop interviewees reported:

> A critical mechanism of knowledge transfer in the district are the leading firms that orchestrate networks of SMEs that produce footwear components. As there is a quite hierarchical structure, SMEs learn primarily form those leading firms that train them to be more efficient and thus improve the whole network.

Finally, in the Toy-Valley district cooperation among competitors is less frequent than in the ceramic tile district, and significant cooperation is observed through the division of labour (producer-user). As members of AIJU reported:

> Cooperation between suppliers-users is very intense, but not among competitors. Imitation, however, is capital in this territory.
>
> Social ties and informal networks really diffuse knowledge around focal local technologies.

4. Industry 4.0 innovation policies for Industry 4.0 in districts

As reported in the reviewed cases, in all the three districts, the regional policymaker (IVACE, belonging to the Economic branch of the Valencian Regional Government) launched an initiative to digitize SMEs ('Agenda Industria 4.0'[8]) in the region. In particular, the principal actors for leading the co-design and implementing the initiative were the different Research and Transfer Institutes operating in each territory. In the ceramic tile district, the project 'CEBRA+' was co-designed with the 'ITC' and the local business association in ceramics (ASCER); in the footwear district, the 'I4FOOTWEAR' was arranged by 'Inescop'; and, in the Toy-Plastic district, the 'TALLER4.0' was co-organized with 'AIJU'.

According to the reviewed literature the sequence of learning aimed at signalling change to SMEs is consistently developed in all the districts: (a) Collective actors (e.g. Research and Transfer Insitutues) access to new digital technology through R&D and open innovation (R&D for digitizing, technology gatekeeping, technology watch, roadmapping, European Union R&D projects, alliances with universities, learning with the information and technology industry, etc.); (b) Collective actors co-design the innovation initiative with public and private local stakeholders in a place-based approach (bottom-up, private and public collaboration, local actors, etc.); (c) The initiative is tailored to the local learning and institutional setting, identifying the main actors to diffuse and contaminate the territory with the new technologies; (d) The collective initiative signals change, shows a roadmap to start from; (e) Diffusion and contamination of the new technology utilizes the district-specific learning and institutional structure, fostering change through imitation and interaction[9]: either diffusing knowledge to leading incumbents that subsequently will 'contaminate' their networks of SMEs (interaction) and/or fostering imitation through managing living-labs to spread imitation. See Table 3 for a more detailed summary of facts.

4.1. The Vinalopo footwear district: 'I4FOOTWEAR' initiative[10]

In the case of the footwear district, the division of labour is very fine-grained and decomposed into multiple parts, from design to leather cutting and sewing but also producing components such as soles, heels, textile parts, plastic parts and many more. Each subsector is very specialized along a multi-step process. Only large or medium-size firms manufacture or coordinate the assembling of the final product (footwear) and those are the ones organizing local suppliers of many footwear components. Therefore, the structure of the networks combines cooperation but also shows a quasi-hierarchical structure where large/medium firms orchestrate networks of SMEs. As reported in Hervás-Oliver (2022), 'Inescop' developed the technology to digitize (CAD/CAM design and 3D printing for prototyping) and primarily targets those firms that orchestrate SMEs to accomplish dissemination of these new digital technologies. This Research and Transfer Institute transfers digital technologies to those leading firms and then, these firms interact with their networks of SMEs to diffuse the new technologies. This is reported to be the learning process in the focal district. Interviewees from Inescop commented (interviews in 2021):

> Small firms cannot absorb new digital technologies, except when they are technology-based startups. These small firms, very specialized in some particular stages of the footwear production, only rely on their customers that outsource them with very precise instructions regarding materials, design and technology. Therefore, we transfer primarily to the leading firms and then, inter-firm interaction diffuses digital technologies to the SMEs.

Table 3. Innovation policies to digitize districts: footwear, plastics-toys and ceramics MIDs.

Dimensions	Footwear MID	Plastics-toys MID	Ceramics MID
Automation and production processes	Labour-intensive, low level of automation.	Medium-low level of automation	High-automation level
Industry 4.0 enablers	Lead and developed by the transfer institute 'Inescop'; 'I4FOOTWEAR' project: CAD-CAM design and 3D printing, among others. Focus on design and marketing activities; less in manufacturing	Lead and developed by the transfer institute 'AIJU'; 'TALLER4.0' project: Augmented reality, 3D printing, Big Data, Robots, others. Focus on manufacturing activities	Lead and developed by the transfer institute 'ITC'; 'CEBRA +' project Sensoring and tracking; Big Data, Robots, others; in two main projects, 'CEBRA' for digitizing manufacturing and 'CEBRA+' for digitizing distribution and sales through on-site digital kiosks for digital display of the product and data collection. Focus on manufacturing activities.
Goal of the policy initiative	Foster change in the district, showing the way to firms; avoid cognitive inertia; facilitating to IT industry the entrance into the industry	Foster change in the district, showing the way to firms; avoid cognitive inertia; facilitating to IT industry the entrance into the industry	Foster change in the district, showing the way to firms; avoid cognitive inertia; facilitating to IT industry the entrance into the industry
Collective initiative in action	Industry 4.0 technology demonstration platform	Industry 4.0 Living-lab demonstration platform at local SMEs	Industry 4.0 Living-lab demonstration platform at a local SME (CEBRA project) Also CEBRA+ digitization of the distribution channel through digital kiosks.
Diffusion mechanism	Utilizing **leading firms** as recipients of the digital knowledge; then, indirect effects (subsequent diffusion and also imitation).	Utilizing **SMEs** (around 50 employees) to place the living-lab for fostering imitation by other local firms.	Utilizing **SMEs** (200-250 employees) to place the living-lab for fostering imitation by other local firms.

Source: Own from literature analysis and empirical evidence (Hervas-Oliver et al. 2019; Hervás-Oliver 2021; Hervás-Oliver 2022), along with subsequent interviews.

> These leading firms orchestrate the networks of SMEs. They design the product and then outsource the production to local SMEs and control it. When these leading firms adopt new digital design technologies, subsequent outsourcing adds new requirements and additional cooperation with their SMEs that are required to learn and utilize new technologies. In the end, leading firms spread the new technology.
>
> We show leading firms the way to digitization. Then, they teach SMEs and also pass down the new way of doing things. It works this way in this district.
>
> Most leading firms work with us and they utilize the new CAD/CAM technologies for footwear, requiring their local SMEs to start to change to the new requirements. It is a very effective diffusion that only works (here) this way.

In this particular case, as reported by Hervás-Oliver (2022), the local cognitive structure showed quasi-hierarchical networks orchestrated by leading firms that coordinate production. SMEs are not really using Research and Transfer Institutes but leading firms do (as they also conduct R&D). This way, the activation of knowledge diffusion in the territory works this way: the Research and Transfer Institute transfers to leading firms that also diffuse to the local networks of SMEs. Inescop has developed a demonstration platform for digitizing and the leading firms use it and engage. It is like a double step process in order to create awareness of change among local SMEs (indirectly through the leading firms that orchestrate local networks).

4.2. The ceramic tile district: CEBRA and CEBRA+

The Ceramic tile district, however, is very different to the footwear district. In accordance with Hervas-Oliver et al. (2019), in the ceramic tile district most firms produce the final product (ceramic tile), albeit there is an ample division of labour (clay extraction, chemical products for decoration, etc.), but the ceramic tile is not decomposable as in the previous case of footwear. Also, the ceramic process is highly automated. Rather, the 'ITC' seeks more to diffuse the new technology by seeking local imitation, rather than using leading firms diffusing within their local networks. As Hervas-Oliver et al. (2019) shows, the 'ITC', along with ASCER (local business association) chose one advanced SME to install a demonstrative living-lab of Industry 4.0 activities. The specific activities for digitizing were primarily sensors and traceability, including data analytics and others. The 'ITC' supported the trial-and-error optimization of the new demonstrative line of production, and then organized open-doors shows to foster subsequent imitation by local firms. Open doors are a capital feature of the initiative, as public (pre-arranged) visits to the living-lab open to local competitors were required. This is possible because of the intense cooperation in the district, including the fact that the local business association considered the initiative as a collective action for the district, to the extent that it chose the company (named 'Colorker') for the living-lab. As interviewees from 'ITC' stated:

> The living-lab demonstration line for Industry 4.0 (CEBRA project) is a way to show the way to local firms, supporting them to start to change.
>
> We are also training the equipment suppliers for ceramic tile machinery to show them how to digitize local ceramic tile firms. We expect that dissemination of best practices by equipment suppliers and imitation in the district will push diffusing new technologies.

After the implementation of this technology demonstration platform, the 'ITC' performed more activities to support change in the district, such as a digital catalogue of new processes to the local equipment auxiliary industry, training activities for digitization, as well as developing a 'simulator' of ceramic tile digitization to show in real terms the expected outcomes. All these activities were focused on ceramic tile production. In addition, the CEBRA+ project started a digitization of sales, creating an on-store 'digital kiosk', as concept, for digitizing sales in physical stores. The new machine, based on Artificial Intelligence, Virtual Reality (immersion technologies) and Machine Learning allows showing real ambiences and spaces where the ceramic tile fits and also collects data from the customers. In this pilot project, 3 local ceramic tile firms ('Azteca, Colorker and Fustecma') and also one store participated.[11] Instead of production, this extension of digitizing is applied to ceramic tile marketing departments.

In this particular case, as Hervas-Oliver et al. (2019) report, the configuration of the local cognitive structure and the local networks favour more diffusion by imitation. Just the opposite to the Footwear initiative, the CEBRA project wanted to utilize SMEs to signal awareness to change, showing the way and to seek pervasive imitation. This differs remarkably from the Footwear diffusion system (through leading firms). In addition, the social ties and extensive cooperation in the territory (e.g. joint purchasing is the norm for gas and utilities, the local business association organizes the largest ceramic tile fair trade in Europe, named CEVISAMA, etc.) facilitate the 'open-doors' policy, where local firms (competitors) could visit the local living-lab at 'Colorker' SME in order to learn the way to change. Imitation is working, as observed by the interviewees.

> It is a very competitive advantage to show the way to digitize to local firms, we did not find these types of initiatives in the industry in any other country.

4.3. The Toy-Valley district: 'TALLER 4.0 initiative'

Lastly, in the Toy-Plastics district, network structure is more socially and relational-based than footwear. In particular, it displays more relational-based networks, similar to ceramics, and most companies produce the final product.

Focusing on the specific local cognitive structure, 'AIJU' undertakes a similar approach to that of the ceramic tile district, seeking imitation and co-designing a living-lab for Industry 4.0 technology demonstration in a SME (the 'Vicedo Marti' firm) with the purpose of developing plastic process oriented Industry 4.0 technology. The latter includes, primarily, tools for developing Artificial Vision, 3D Printing, Augmented Reality and Data Analytics applied to plastics, among other technologies. It also follows an 'open-doors' approach, as stated by Hervás-Oliver (2021). This means that local competitors can visit the digitized production line to see how to do it and what the outcome could be. This initiative signals change and spread 'me too' or imitation among local SMEs in that industry. In this case, the substantial difference with the ceramic tile district is primarily based on the fact that firms are rather smaller than the ceramic ones, meaning less resources to digitize. In this case the advanced SME (size around 50 employees) chosen for the living-lab is very different from the one chosen in the ceramic tile district (size over 200 employees). As interviewees in 'AIJU' pointed out:

> It was not our intention to choose the best or most advanced firms, but one that can inspire other SMEs to follow suit. Our purpose is to prepare the district for change and show the way but also to give the message: all firms can do it.

> We expect imitation in the district, as the firm is a reference point for other SMEs

> Large firms in the district have their R&D funds for digitizing and their headquarters' support. These firms are not within our realm. Rather, SMEs are the core of the district and our purpose.

4.4. General insights

In all cases, all the Research and Transfer Institutes manifested the crucial importance of the co-design of the innovation policy instrument, interacting both formal and informally with local business associations, local companies and the public policymakers (named IVACE). All in all, those initiatives are built upon associative structures of governance that are bottom-up, decentralized, open, consultative and involve local actor coordination, allowing co-design by public and private stakeholders that present a common understanding about the territory and its change (e.g. Feldman and Lowe

2018). In addition, each place followed a specific place-based initiative. There is no point in creating a single one-size-fits-all initiative for different places.

As 'AIJU' indicated:

> We visited the Ceramic tile (CEBRA) living-lab, nice as an inspiration, but there is no point in creating a single one-size-fits-all initiative for different realities. Our district is very different in all dimensions.
>
> We co-designed the initiative, bringing our vision to the negotiation with policymakers. They trust us, we have plenty of experience in the territory
>
> This bottom-up approach is the best for the territory, as it also decides what to do from consultation with the real actors: local firms and local support organizations
>
> The Taller 4.0 created awareness, showed the way to SMEs, achieved imitation and created a new IT industry specialized in the territory and its needs.

As 'Inescop' stated:

> Policymakers just set the goal (digitization) and we proposed the best possible way for the territory, considering many different factors such as the local technology, the type of networks and other factors. We explained it and they agreed, setting some boundary conditions. This co-creation is very fruitful, as the local place is put first.
>
> We adapt the Industry 4.0 initiative for our local system: leading firms organize production and innovation. Diffusion through leading firms is how learning occurs in this district
>
> Also imitation is important, as everybody knows each other. What leading firms do is the next step for SMEs.
>
> We did not choose a living-lab, because cooperation mainly occurs in the user-producer interaction and because there are many different activities in the local value chain … ..diffusion based on leading-SMEs interaction has always worked in here … ..
>
> I4FOOTWEAR has proved to be successful, most of the leading firms have initiated the change and utilize our technology. The spread it throughout the territory.

These type of indirect effects on SMEs, are evidenced as non-targeted SMEs might improve their performance when they are in the networks (as suppliers or customers) of target leading firms that receive the knowledge from the universities or research centres (see Fotso 2022; Chai and Shih 2016).

As 'ITC' reported:

> Allowing the local organizations to lead the change is the best way, as we are permanently in touch with the territory and know its technology and needs. Thus, we negotiated with policymakers and the local business association (ASCER) what we should do and how to do it, and after some changes and refinements, all actors agreed on the plan.
>
> CEBRA is the first digitization initiative for the ceramic tile industry founded on a living-lab and open doors basis, to the extent that many international competitors were interested to visit the plant, from Brazil, Mexico, Italy, etc.
>
> When policymakers allow local actors to show the way, it works much better as it is not something imposed.

> Each local company is going to use CEBRA in its own way, as each of them have different capabilities. We wanted mainly to show the way and to show that is possible to do it, for this reason we chose an SME.
>
> We went further and also proposed the CEBRA+, adding digitization of sales, through digital kiosks, and going beyond the initial idea of just digitizing production. This was a need from the local firms that was not previously considered by policymakers who were just thinking about production improvements.
>
> The territory is changing and most companies are investing in digitization, as we achieve the generation of best practices and legitimization of the new paradigm. Nevertheless, awareness is larger than adoption at this moment.

The policy instrument for all districts was also supported by an innovation voucher scheme for incentivizing SMEs to start to adopt Industry 4.0 ('DigitalizaCV' programme, launched by policymaker named IVACE), training courses, demonstration events, seminars, etc.

In each MID, the policy was a place-based effort where local stakeholders (especially RTIs but also some local business associations) negotiated what to do and how to do it. Notice that the local Research and Transfer Institutes are organized by local businessmen, researchers and scientists and policy-makers' representatives.

As reported in the literature, each district initiative was totally tailored to local productive 'chorality' of places (Becattini 2015), that is, place-based and collective in nature: actors, technology, networks-institutions and collective actors.

As regards local actors, we identified important differences across the three MIDs compared, as indicated in the previous Section. Then, for the case of local technologies, there are remarkable differences not only in the type of technologies but their level of automation in each district. While the footwear district is still a very labour-intensive industry, the toy-plastic is intermediate and the ceramic tile district is almost totally automated. Thus, automation facilitates digitization, as most digitization processes consists of adding new sensors to existing automated processes. The type of local networks and their structure in the district, i.e. how the division of labour is organized and how the final product is accomplished, is also very important. It describes not only how value is added, but how the learning process occurs and how collaboration is organized.

5. Conclusions

This study approaches innovation policies for Industry 4.0 in MIDs by analysing collective initiatives undertaken in different industrial districts. In this context, we answer the following research questions that constitute this study's goal: 'what' types of innovation policies work for Industry 4.0 in districts? And, where they do, 'how' are those collective innovation initiatives designed, developed and implemented? We assume that SMEs in districts struggle to adopt radical changes and they also present the typical problems of SMEs (e.g. poor financial resources and competences, weak information and technology systems, etc.). For this reasons, initiatives aimed at collectively signalling change towards Industry 4. 0 and showing how to do it are very interesting: collective and cooperative place-based initiatives.

By reviewing existing literature and performing additional empirics (24 interviews in 2021) on three industrial districts in Spain (i.e. the Vinalopo footwear district, the Toy-Valley district and the Ceramic tile district of Castellon, all in the Valencian Region, the one with more districts in Spain, see Boix and Galletto 2009), we are able to build theory inductively by comparing each territory's collective initiative for Industry 4.0. Results reveal that the same Industry 4.0 programme led to different policy initiatives or instruments by involving local actors' co-design, negotiation and implementation: these collective actors, mainly RTIs and local business associations, tailored each initiative to their focal territories and idiosyncratic characteristics.

Industry 4.0 policies in these settings were place-based: co-designed, openly discussed, tailored and implemented in a very local-sensitive approach. Findings suggest that the relevant factors of these policies are:

First, the composition of local firms and their characteristics, especially size, organizational strategies, technology, etc., matters: it really makes a difference local SME capabilities (e.g. finance, information systems, managerial, etc.) and their industries that require different digital enablers.

> Proposition 1: Typologies of local SME capabilities and industries agglomerated in the focal district influence Industry 4.0 initiatives.

Second, the structure and function of institutions and local networks, especially defining the local production, innovation and learning process: each territory presents an idiosyncratic 'way of learning'. As we consider that both learning from (imitation) and learning with (interaction) 'à la Staber' are very important in districts, this dimension really influences diffusion of the new digital technologies. Thus, open doors policies (and their open living-labs), based on intense social capital, differs in each of the three districts analysed.

> Proposition 2: Local institutions and networks interaction and learning process in the focal district influence Industry 4.0 initiatives.

Third, collective actors lead and organize the cooperative collective action. Local collective actors (local business associations, Research and Transfer Institutes, etc.) are very important in these policies, as they mediate the public and private local actors, share a collective understanding of local needs and also legitimate the new Industry 4.0 technology for being accepted in the territory as a new sub-identity. They signal change and avoid cognitive inertia, especially among SMEs. Also, collective actors utilize the cooperation–competition logic of MIDs, aligning interests and coordinating all actors required for the change (new companies from information system industries, new complementary activities, etc.).

> Proposition 3: Local collective actors shape the Industry 4.0 initiative utilizing the cooperation-competition logic of MIDs, legitimizing change and aligning common interests of the focal districts.

One of the main differences between regional and national policies for Industry 4.0 is the place-based approach of the former (see more at Hervas-Oliver et al. 2021): tailored to the local context, coordinated and co-designed with local actors and promoting change but considering the local technology and institutional setting. Space-blinded national policies are interesting for changing skills and capabilities across industries (e.g. training in digital technologies, financial schemes for adopting Industry 4.0) but do not understand the collective and competition–cooperation idiosyncratic nature of districts, nor their local networks. Hence, national ones are a good complement to the regional ones that are the core ones. Coordination, nevertheless, should be considered from regional policies to avoid overlapping.

The study's results contribute to the Marshallian literature (e.g. Belussi and Sedita 2009; Hervás Oliver 2015; Hervas-Oliver, Lleo, and Cervello 2017) and the Industry 4.0 topic (e.g. Bellandi, de Propris, and Santini 2019; Bellandi et al. 2020; Hervas-Oliver 2021; Lazzeretti et al. 2022).

Notes

1. See more on the differences between digitization and digitalization (e.g. Yoo et al. 2012).
2. https://s3platform.jrc.ec.europa.eu/digital-innovation-hubs.
3. https://www.gov.uk/government/publications/industrial-strategy-building-a-britain-fit-for-the-future.
4. https://catapult.org.uk/.
5. Department for Business, Energy & Industrial Strategy (2017). Catapult Program: A Framework for Evaluating Impact. Retrieved from: https://assets.publishing.service.gov.uk/government/uploads/system/uploads/attachment_data/file/662319/catapult-programme-evaluation-framework.docx.pdf.
6. From local endogenous and incremental learning, Mark III category goes to more show district openness to connect to trans-local and trans-international value chains, *servitization*, multinational companies and more radical changes. See more in Bellandi's works.
7. https://www.redit.es/en/home/
   ITC: Ceramics Technology Institute; Inescop: Footwear Technology Institute; AIJU: Toys and Plastic Technology Institute.
8. https://www.ivace.es/images/Industria_4.0/Agenda_Industria_40_CV_v0_web.pdf.
9. See Staber (2009).
10. Based on Hervás-Oliver (2022).
11. See more here: https://ceria.es/category/itc/.

Disclosure statement

No potential conflict of interest was reported by the author(s).

Funding

Financial support from AICO2020/123 (*Consolidables*, Generalitat Valenciana); RTI2018-095739-B-100 (Ministerio de Ciencia e Innovación, Gobierno de España, AEI/EU FEDER); PID2021-128878NB-I00 (Ministerio de Ciencia e Innovación, Gobierno de España, AEI/EU FEDER). Also, Dr. Hervas-Oliver acknowledges funding from Generalitat Valenciana, for visiting research period at *University of Padova*, 2022 (CIBEST/2021/238).

ORCID

Jose-Luis Hervas-Oliver http://orcid.org/0000-0001-6438-5497
Jose Antonio Belso-Martínez http://orcid.org/0000-0003-2985-7481

References

Barca, F., P. McCann, and A. Rodríguez-Pose. 2012. "The Case for Regional Development Intervention: Place-Based Versus Place-Neutral Approaches." *Journal of Regional Science* 52 (1): 134–152. doi:10.1111/j.1467-9787.2011.00756.x

Becattini, G. 2015. "Beyond Geo-Sectoriality: The Productive Chorality of Places." *Investigaciones Regionales – Journal of Regional Research* 32 (2015): 31–41.

Bellandi, M., L. de Propris, and E. Santini. 2019. "Industry 4.0+ Challenges to Local Productive Systems and Place-Based Integrated Industrial Policies." In *Transforming Industrial Policy for the Digital Age*, 201–218. Edward Elgar Publishing. doi:10.4337/9781788976152.00016.

Bellandi, M., E. Santini, C. Vecciolini, and L. De Propris. 2020. "Industry 4.0: Transforming Local Productive Systems in the Tuscany Region." In Industry 4.0 and Regional Transformations, 84–96. Routledge.

Belussi, F., and S. Sedita. 2009. "Life Cycle vs. Multiple Path Dependency in Industrial Districts." *European Planning Studies* 17 (4): 505–528. doi:10.1080/09654310802682065

Bettiol, M., M. Capestro, E. Di Maria, and S. Micelli. 2021. "Disentangling the Link Between ICT and Industry 4.0: Impacts on Knowledge-Related Performance." *International Journal of Productivity and Performance Management* 71 (4): 1076–1098.

Boix, R., and V. Galletto. 2009. "Innovation and Industrial Districts: A First Approach to the Measurement and Determinants of the I-District Effect." *Regional Studies* 43 (9): 1117–1133. doi:10.1080/00343400801932342

Boix, R., V. Galletto, and F. Sforzi. 2019. "Place-based Innovation in Industrial Districts: The Long-Term Evolution of the iMID Effect in Spain (1991–2014)." *European Planning Studies*, 1940–1958. doi:10.1080/09654313.2019.1588861

Chai, S., and W. Shih. 2016. "Bridging Science and Technology Through Academic–Industry Partnerships." *Research Policy* 45 (1): 148–158. doi:10.1016/j.respol.2015.07.007.

Corò, G., and M. Volpe. 2020. "Driving Factors in the Adoption of Industry 4.0 Technologies: An Investigation of SMEs." In *Industry 4.0 and Regional Transformations*, edited by L. De Propris, and D. Baley, chapter 7, 112–132. London: Routledge.

De Propris, L., and D. Bailey. 2020. *Industry 4.0 and Regional Transformations*. London: Taylor & Francis.

Ebbekink, M., and A. Lagendijk. 2013. "What's Next in Researching Cluster Policy: Place-Based Governance for Effective Cluster Policy." *European Planning Studies* 21 (5): 735–753. doi:10.1080/09654313.2013.734460

Eisenhardt, K. M., and M. E. Graebner. 2007. "Theory Building From Cases: Opportunities and Challenges." *Academy of Management Journal* 50 (1): 25–32.

Feldman, M., and N. Lowe. 2018. "Policy and Collective Action in Place." *Cambridge Journal of Regions, Economy and Society* 11 (2): 335–351. doi:10.1093/cjres/rsy011

Fotso, R. 2022. "Evaluating the Indirect Effects of Cluster-Based Innovation Policies: The Case of the Technological Research Institutes in France." *The Journal of Technology Transfer* 47 (4): 1070–1114.

Galletto, V., and R. B. Domènech. 2014. "Industrial Districts, Technological Innovation and I-district Effect: A Question of Volume or Value?" *Investigaciones Regionales/Journal of Regional Research*, no. 30: 27–51.

Gilbert, B. 2012. "Creative Destruction: Identifying its Geographic Origins." *Research Policy* 41: 734–742. doi:10.1016/j.respol.2011.11.005.

Glasmeier, A. 1991. "Technological Discontinuities and Flexible Production Networks: The Case of Switzerland and the World Watch Industry." *Research Policy* 20: 469–485. doi:10.1016/0048-7333(91)90070-7.

Hervás Oliver, J. L. 2015. "How do Multinational Enterprises co-Locate in Industrial Districts? An Introduction to the Integration of Alternative Explanations from International Business and Economic Geography Literatures." *Investigaciones Regionales* 32: 115–132.

Hervás-Oliver, J. L. 2021. "Industry 4.0 in Industrial Districts: Regional Innovation Policy for the Toy Valley District in Spain." *Regional Studies* 55 (10–11): 1775–1786. doi:10.1080/00343404.2021.1939861

Hervás-Oliver, J. L. 2022. "Industry 4.0 in the Vinalopo Footwear District: The Leading Role of Supporting Organizations for Transformation." *Competitiveness Review*, 1059–5422. doi:10.1108/CR-06-2022-0075.

Hervás-Oliver, J., L. J. Albors-Garrigos, S. Estelles-Miguel, and C. Boronat-Moll. 2018. "Radical Innovation in Marshallian Industrial Districts." *Regional Studies* 52 (10): 1388–1397. doi:10.1080/00343404.2017.1390311

Hervas-Oliver, J. L., J. A. Belso-Martínez, and I. Díez-Vial. 2022. "Multinationals' Recruiting in Industrial Districts." *Regional Studies* 56 (8): 1320–1332.

Hervas-Oliver, J., S. Estelles-Miguel, G. Mallol-Gasch, and J. Boix-Palomero. 2019. "A Place-Based Policy for Promoting Industry 4.0: The Case of the Castellon Ceramic Tile District." *European Planning Studies* 27 (9): 1838–1856. doi:10.1080/09654313.2019.1642855

Hervas-Oliver, J., L. G. Gonzalez-Alcaide, R. Rojas-Alvarado, and S. Monto-Mompo. 2021. "Emerging Regional Innovation Policies for Industry 4.0: Analyzing the Digital Innovation hub Program in European Regions." *Competitiveness Review: An International Business Journal* 31 (1): 106–112.

Hervas-Oliver, J., L. G. Gonzalez, P. Caja, and F. Sempere-Ripoll. 2015. "Clusters and Industrial Districts: Where is the Literature Going? Identifying Emerging Sub-fields of Research." *European Planning Studies* 23 (9): 1827–1872. doi:10.1080/09654313.2015.1021300

Hervas-Oliver, J. L., M. Lleo, and R. Cervello. 2017. "The Dynamics of Cluster Entrepreneurship: Knowledge Legacy From Parents or Agglomeration Effects? The Case of the Castellon Ceramic Tile District." *Research Policy* 46 (1): 73–92.

Lazerson, M., and G. Lorenzoni. 1999. "The Firms That Feed Industrial Districts: A Return to the Italian Source." *Industrial and Corporate Change* 8: 235–266. doi:10.1093/icc/8.2.235.

Lazzeretti, L., N. Innocenti, M. Nanelli, and S. Oliva. 2022. "The Emergence of Artificial Intelligence in the Regional Sciences: A Literature Review." *European Planning Studies*, doi:10.1080/09654313.2022.2101880.

Lepore, Dominique, and Francesca Spigarelli. 2020. "Integrating Industry 4.0 Plans Into Regional Innovation Strategies." *Local Economy: The Journal of the Local Economy Policy Unit* 35 (5): 496–510. doi:10.1177/0269094220937452

Magro, E., and J. Wilson. 2019. "Policy-mix Evaluation: Governance Challenges from new Place-Based Innovation Policies." *Research Policy* 48: 103612. doi:10.1016/j.respol.2018.06.010.

Matt, D., M. Molinaro, G. Orzes, and G. Pedrini. 2021. "The Role of Innovation Ecosystems in Industry 4.0 Adoption." *Journal of Manufacturing Technology Management* 32 (9): 369–395.

Moeuf, A., S. Lamouri, R. Pellerin, S. Tamayo-Giraldo, E. Tobon-Valencia, and R. Eburdy. 2020. "Identification of Critical Success Factors, Risks and Opportunities of Industry 4.0 in SMEs." *International Journal of Production Research* 58: 1384–1400. doi:10.1080/00207543.2019.1636323.

Müller, J., O. Buliga, and K. Voigt. 2018. "Fortune Favors the Prepared: How SMEs Approach Business Model Innovations in Industry 4.0." *Technological Forecasting and Social Change* 132: 2–17. doi:10.1016/j.techfore.2017.12.019.

Munari, F., M. Sobrero, and A. Malipiero. 2012. "Absorptive Capacity and Localized Spillovers: Focal Firms as Technological Gatekeepers in Industrial Districts." *Industrial and Corporate Change* 21: 429–462. doi:10.1093/icc/dtr053.

Oesterreich, T., and F. Teuteberg. 2016. "Understanding the Implications of Digitisation and Automation in the Context of Industry 4.0: A Triangulation Approach and Elements of a Research Agenda for the Construction Industry." *Computers in Industry* 83: 121–139. doi:10.1016/j.compind.2016.09.006

Pagano, A., E. Carloni, S. Galvani, and R. Bocconcelli. 2021. "The Dissemination Mechanisms of Industry 4.0 Knowledge in Traditional Industrial Districts: Evidence from Italy." *Competitiveness Review*, doi:10.1108/CR-12-2019-0160.

Porter, M., and J. Heppelmann. 2014. "How Smart, Connected Products are Transforming Competition." In *Harvard Business Review* (Issue November 2014).

Sandulli, F., E. Gimenez-Fernandez, and M. Rodriguez Ferradas. 2021. "The Transition of Regional Innovation Systems to Industry 4.0: The Case of Basque Country and Catalonia." *European Planning Studies* 29 (9): 1622–1636. doi:10.1080/09654313.2021.1963049

Sommer, L. 2015. "Industrial Revolution – Industry 4.0: Are German Manufacturing SMEs the First Victims of This Revolution?" *Journal of Industrial Engineering and Management* 8 (5): 1512–1532. doi:10.3926/jiem.1470.

Staber, U. 2009. "Collective Learning in Clusters: Mechanisms and Biases." *Entrepreneurship and Regional Development* 21: 553–573. doi:10.1080/08985620802529526.

Teixeira, J., and A. Tavares-Lehmann. 2022. "Industry 4.0 in the European Union: Policies and National Strategies." *Technological Forecasting and Social Change* 180: 121664. doi:10.1016/j.techfore.2022.121664

Uotila, T., V. Harmaakorpi, and R. Hermans. 2012. "Finnish Mosaic of Regional Innovation System—Assessment of Thematic Regional Innovation Platforms Based on Related Variety." *European Planning Studies* 20 (10): 1583–1602. doi:10.1080/09654313.2012.713331

Yoo, Y., R. Boland, Jr., K. Lyytinen, and A. Majchrzak. 2012. "Organizing for Innovation in the Digitized World." *Organization Science* 23 (5): 1398–1408. doi:10.1287/orsc.1120.0771.

York, J., T. Hargrave, and D. Pacheco. 2016. "Converging Winds: Logic Hybridization in the Colorado Wind Energy Field." *Academy of Management Journal* 59, doi:10.5465/amj.2013.0657.

Yüksel, H. 2020. "An Empirical Evaluation of Industry 4.0 Applications of Companies in Turkey: The Case of a Developing Country." *Technology in Society* 63: 101364. doi:10.1016/j.techsoc.2020.101364

The actors' role in industrial districts facing the challenge of technologies 4.0 between absorption and dissemination

Niccolò Fiorini, Tommaso Pucci, Elena Casprini and Lorenzo Zanni

ABSTRACT

Industry 4.0 (I.4.0) has a significant impact on firms, supply chains and in contexts where firms have built tight relations with high levels of interactions in a delimited geographical area. An industrial district (ID) represents an extraordinary field for empirical investigation to understand how the knowledge linked to the I.4.0 technological paradigm can be absorbed and disseminated. However, the current literature has focused its attention on the effect of I.4.0 on the IDs at an aggregate level, while little is known about the role of the firms within the IDs. This research aims to fill this gap by looking at two dimensions: absorption and dissemination. We focused our attention on 101 firms belonging to four Tuscan IDs: the Santa Croce tanning district, the Florentine leather district, the Prato textile district and the Arezzo jewellery district. This research broadens the literature about absorption and dissemination within IDs enlightening the behaviour of firms in IDs, where they can play different roles: 'traditionalist', 'bridge', 'isolated' and 'opportunist'. From a practical perspective, this paper shows firms their role played within IDs. From this research it comes out that competencies 4.0, culture 4.0 and collaboration are the key aspects for turning IDs into IDs 4.0.

1. Introduction

Industry 4.0 (I.4.0) is leading to a disruptive effect in any industry and in most places of the world (Xu, David, and Kim 2018). Among its significant aspects, we must include digital integration, which takes place among physical and non-physical structures and crosswise organizational borders (Schumacher, Erol, and Sihn 2016). I.4.0 technologies affect the way firms are creating and exchanging value, as well as how they are collaborating with each other (Bibby and Dehe 2018; Ilvonen et al. 2018; Kipper et al. 2020).

Therefore, the role of I.4.0 technologies could be particularly relevant in contexts where firms are traditionally linked, such as in the case of local economic systems as Industrial Districts (IDs). In such contexts, firms are exposed to higher levels of interaction due to geographical proximity and long-lasting relationships, thus being in a privileged position to both absorb knowledge (Cohen and Levinthal 1990) and disseminate it (see Owen-Smith and Powell 2004). Considering this, I.4.0 and digitalization can exert a pivotal role both in promoting the exchange of knowledge within the same spatial agglomeration of firms and in facilitating the import of knowledge from outside.

However, some preliminary research has been conducted in analysing how IDs affect I.4.0 implementation (Grashof et al. 2020), promote the I.4.0 path (Götz and Jankowska 2017), help managing I.4.0 (Bellandi and De Propris 2017), little is known about the different roles that firms may play within an ID regarding how they acquire, manage and use the new technological knowledge. Some scholars have studied the role of certain actors within firms' spatial agglomeration focusing on their interactions leading to innovation (Guercini and Runfola 2015; Lazzeretti and Capone 2016; Pucci et al. 2018; Innocenti, Capone, and Lazzeretti 2020). Even if literature analysing districts and innovation has increased over the time (Caloffi, Lazzeretti, and Sedita 2018), the effects of I.4.0 on the interaction mechanisms within an ID are still scarcely investigated. Since a new paradigm as I.4.0 could change the (already studied) ways firms interact within IDs, this paper aims to investigate the role of players within IDs with respect to I.4.0. It looks at two dimensions to describe how firms may behave within an ID. The first one is 'absorption', referring to the firm's capacity to assimilate external knowledge (Cohen and Levinthal 1990; Nooteboom 2000). The second dimension is 'dissemination': within local economic systems, some firms may act as gatekeepers (Malipiero, Munari, and Sobrero 2005; Morrison 2008) that not only absorb, but also make knowledge spilling over the local context. Throughout dissemination, knowledge is diffused among several actors, thus boosting cross-fertilization and innovation (Lazzeretti and Capone 2009; Lazzeretti and Capone 2015). Hence, this work aims to understand the degree of absorption and dissemination with respect to I.4.0 in firms belonging to IDs. Furthermore, we want to investigate which role those firms exert in IDs. To accomplish with these goals, the paper is organized as follows: first, we analyse the theoretical background and we develop the conceptual model for the analysis; then, we describe the methodology; results are reported thereafter, with the conclusion and limitations closing the paper.

2. Theoretical background and framework

The origin of the term I.4.0 arose from the Hannover Fair, 2011, in Germany with the term Industrie 4.0. The aim of the German nationwide project was to push the digitalization of the national manufacturing industry (Kagermann, Wahlster, and Helbig 2013). After this first German program, several other countries, not only in Europe (Piccarozzi, Aquilani, and Gatti 2018; Álvarez Marcos, Capelo Hernández, and Álvarez Ortiz 2019), but worldwide (Li 2018; Takakuwa, Veza, and Celar 2018) started to design their own I.4.0 agenda. Therefore, a peculiar characteristic of I.4.0 is that it was a 'planned' industrial revolution (Lasi et al. 2014): researchers, experts and practitioners named it a revolution before it started, while the revolutionary impact of the previous three was awarded later. However, Kagermann (2015) states that the impact of I.4.0 regarding the economic development and the organization of work has the same level of intensity as the previous three revolutions. Furthermore, scholars agree in defining I.4.0 as a paradigm. Even if someone might perceive the whole process as a natural evolution of something already existent (i.e. simply an evolution of previous technologies), it is important to underline that the approach used is completely new (Rojko 2017) and this represents the key aspect of I.4.0. I.4.0 presents unprecedented features such as velocity, scope and system impact (Xu, David, and Kim 2018). Thus, the fourth one has the same dignity as all the other industrial revolutions, and it is not considered an industry itself, but rather a set of enabling technologies that can be implemented, and that has disruptive effects, in all industries and in both developed and developing countries.

Given that I.4.0 can legitimately be considered the fourth industrial revolution, the subject deserves more theoretical and empirical in-depth analysis. Several works concentrate their effort on analysing the enabling technologies, such as additive manufacturing, 3D printing, augmented reality, big data and analytics, advanced manufacturing solutions, cyber-security, cloud, blockchain, industrial internet, internet of things, horizontal/vertical integration, smart manufacturing and simulation (Fantoni et al. 2017b; Majeed and Rupasinghe 2017). Others analyse managerial impacts rather than single technologies, e.g. logistics (Winkelhaus and Grosse 2020), e-business (Kulyk and Škodová Parmová 2017), digitalization and value creation (Vasin et al. 2018; Garzoni et al. 2020), impacts on production processes (Ghobakhloo 2018), optimization (Schuh et al. 2014), maintenance services (Suzuki et al. 2014) and supply chain management (Witkowski 2017). Few academics also enlighten the enabling effect of I.4.0 in terms of product-service-development (Kaltenbach et al. 2018).

2.1. Knowledge management dynamics and peculiarities in industry 4.0 framework

Due to the complexity of understanding I.4.0 technologies, firms are often uncertain about whether and, in particular, how to adopt and implement them (Bibby and Dehe 2018; Bienhaus and Haddud 2018). In fact, the adoption of I.4.0 technologies creates new ways for communicating, exchanging data, integrating systems and chains, together with an enhanced flexibility and efficiency (Rüßmann et al. 2015; Schumacher, Erol, and Sihn 2016; Ding 2018): I.4.0 is not restricted to the production processes, but it is a new paradigm, an innovative way that has effects on all the processes and behaviours of firms, changing their interactions both internally and externally. Indeed, a crucial aspect of I.4.0 lies on the approach towards technologies rather than on the implementation of technologies itself (Rojko 2017), which furthermore are not even completely new, and hence the pivotal novelty is the way the involved technologies are capitalized (Baur and Wee 2015), pushing to a completely new domain (Almada-Lobo 2015).

The technological innovation led by I.4.0 is undisputable, however the effects on firms' behaviour and organizations (both internal and external) are relevant. For example, processes are accelerated and more efficient thanks to I.4.0 (Dalenogare et al. 2018; Xu, David, and Kim 2018). Furthermore, flexibility allows new managerial tools like lean manufacturing (Rüttimann and Stöckli 2016), changing the firms' processes internally and, at the same time, modifying their needs from other players. Moreover, we could list several other not production-oriented effects of I.4.0, such as transparency and the increased importance of modularity. Transparency is possible for the whole chain (Fatorachian and Kazemi, 2021) due to the integration of IT with traditional hardware and software (Thames and Schaefer 2016). With respect to modularity, it has been existing far before I.4.0 (Koren and Shpitalni 2010), but with the fourth industrial revolution paradigm the previous rigid borders between firms are feeble and this pushes firms to be modular for taking advantage of this (Pirola, Cimini, and Pinto 2020). All the above-mentioned (not exhaustive) effects of I.4.0 take place as a consequence of both technological implementation and new inter- and intra- organizational approaches and governances due to the new 4.0 paradigm. According to this, knowledge is also spread and acquired differently. The way firms, thanks to I.4.0, might come across new technologies, also from other industries, and circulate them amongst suppliers, customers and partners is different. For example, the prior-mentioned modularity and blur boundaries (Ilvonen et al. 2018; Kohtamäki et al. 2019) make possible to increase the cooperation and the inter-organizational relationships (Halldórsson, Hsuan, and Kotzab 2015; Ericson et al. 2018), hence also increasing the exchange of knowledge among them. Furthermore, intra- and inter-firm collaboration is the key for fully using the potentialities of I.4.0, multiplying the beneficial effects of this revolution (Ilvonen et al. 2018; Kipper et al. 2020). Firms might be forced to adopt some new technologies to cooperate with some partners, strengthening a virtuous circle where not only one firm brings innovation into other firms, but also this empowers extra benefits (Ghadge et al. 2020) for them and the society.

However, large firms and small and medium enterprises (SMEs) do not always behave in the same way towards I.4.0 (Horváth and Szabó 2019). Even if the academic research is more focused on larger firms (Schmidt et al. 2015), we would like to deepen these aspects, since IDs include a significant number of SMEs.

Focusing on how I.4.0 deals with SMEs, some authors underline that SMEs' implementation of I.4.0 technologies is lagging behind (Yu and Schweisfurth 2020) in comparison to big firms. One of the problems of SMEs compared to big ones, is the weaker network connection they have (Mittal et al. 2018). However, firms in IDs can be part of strategic networks (Lazzeretti and Capone 2017).

Indeed, I.4.0 is detrimental for SMEs for maintaining competitiveness, even if in most cases they need more expertise: without implementing I.4.0 they would be soon driven out of the market (Müller, Buliga, and Voigt 2018). In their research, Müller, Buliga, and Voigt (2018) underline the difficulties of SMEs to approach and enable themselves as I.4.0 firms, albeit they are generally aware of the benefits of I.4.0 implementation. Büchi, Cugno, and Castagnoli (2020, 6) verify the existence of a 'causal relationship between their openness to Industry 4.0 and performance'. Even more benefits arise if SMEs do cooperate with partners, increasing their absorptive capacity (Müller, Buliga, and Voigt 2021). Those benefits derived from SMEs' openness could be in terms of, amongst others: higher production capacity, quality, speed, flexibility and also linked to the possibility to exert mass customization and personalization (Büchi, Cugno, and Castagnoli 2020).

2.2. Industry 4.0 within industrial districts

Nevertheless, the way I.4.0 affects the exchange, and the propagation of new technologies, knowledge and capabilities can change according to the ecosystem in which the firm operates. Among others, IDs represent an interesting case to be analysed for understanding the impact of I.4.0. Unfortunately, studies on the effects of I.4.0 on IDs are exiguous and recent. We can start from analysing the literature that, without referring to IDs, emphasizes the I.4.0 effects of horizontal, vertical and end-to-end integration for the supply chain actors (Kagermann, Wahlster, and Helbig 2013), and then we could presuppose that similar effects might appear also in IDs. In fact, even if some scholars might see IDs as too territory-based that could hardly have connections with a paradigm composed by far distance sharing and communication enabling technologies (Alcácer, Cantwell, and Piscitello 2016), its close network relations can provide the perfect environment for developing and implementing I.4.0 (Götz and Jankowska 2017) even if adjustments for a digital transformation and a data-driven world are needed. A transformational digital disruption pushed by I.4.0 can have effects also on IDs (Bellandi, De Propris, and Santini 2019). To analyse such effects, due to the lack of consistent literature about I.4.0 and IDs, some scholars (Hervas-Oliver et al. 2019) propose exploratory studies on how a radical and discontinuous innovation encouraged by I.4.0, which also includes a boundary-spanning perspective, could work and be also fostered by IDs, traditionally oriented to incremental innovation and with limited geographical influence and weak connections outside the IDs. Given that industrial revolutions are the result of technological jumps that cause paradigm shifts (Lasi et al. 2014), also in this fourth industrial revolution we have a paradigm shift. We then need to understand how to analyse the impact of I.4.0 on IDs.

Within IDs, I.4.0 boosts new development paths, recombines the productive knowledge and leads to a collective rerouting (Bellandi et al. 2020). How IDs manage I.4.0 is another question posed in some preliminary studies (Bellandi and De Propris 2017). IDs can benefit from I.4.0 without losing their peculiarities while instead they can fruitfully combine their artisanal distinctiveness together with some I.4.0's characteristics such as the already mentioned flexibility, but also reduced time-to-market, greater product diversification and more customer orientation (Bettiol et al. 2020). For propagating the 4.0 paradigm, leading firms and actors operate as disseminators towards smaller and less technology-opened firms within the cluster (Hervás-Oliver 2021; Pagano et al. 2020). Together with them, some other studies have been recently conducted by scholars for understanding the implementation of I.4.0 in IDs (Grashof et al. 2020). However, more analysis is needed, also because the existent research mainly focuses on the role of IDs in adapting or fostering I.4.0, but from a district's perspective rather than from a firm's perspective within an ID (Hervas-Oliver, Di Maria, and Bettiol 2021).

Due to the lack of a study that focuses on the firms' behaviour within the ID during the fourth industrial revolution, we would like to analyse how firms could assimilate and eventually circulate the paradigm 4.0. For achieving this purpose, we start from studies analysing absorptive capacity and knowledge dissemination (Munari, Sobrero, and Malipiero 2012; Parra-Requena, Ruiz-Ortega, and Garcia-Villaverde 2013) that underline the firms' role within the districts for acquiring and circulating knowledge. For the analysis of the absorptive capacity, an interesting approach is used by Giuliani (2003; and then adapted in 2005). The author analyses the levels of absorptive capacity and the related features as the analytical framework, that fits with the characteristics of a district, as pointed out also in the studies at the basis of our theoretical path (Munari, Sobrero, and Malipiero 2012; Parra-Requena, Ruiz-Ortega, and Garcia-Villaverde 2013). Furthermore, several other authors (Belussi 2005; Camuffo and Grandinetti 2011a; Molina Morales, De Marchi, and Martínez-Cháfer 2021) indicate the importance of absorption and, therefore, of the absorptive capacity for innovation when analysing the diffusion of significantly innovative technologies. Together with this, dissemination (Molina-Morales, Martínez-Cháfer, and Valiente-Bordanova 2017) is considered a crucial aspect for understanding the impact of significant innovations in IDs. Consequently, we would like to build our theoretical framework including absorption and dissemination, instead of deepening the aspects of how a firm identifies, assimilates and exploits knowledge (Cohen and Levinthal 1990), as done by Giuliani (2003). Hence, our aim is to include the degree of firm's absorption and dissemination in an ID, within the I.4.0 paradigm. To do so, we modify the framework proposed by Giuliani (2003) similarly to what Guercini and Runfola (2015) propose. In fact, rather than on the whole district as done by Giuliani (2003), they focus on the individual firm: this perspective is the one we would like to follow.

2.3. Conceptual model and research questions

Simplifying what Giuliani (2003) suggests, we consider the absorption of each firm as, alternatively, high or low. While adapting the matrix proposed by Guercini and Runfola (2015), we then combine the level of absorption of the firm with its dissemination ability, which again could be, alternatively, high or low. The reason why we decide to avoid any specification on whether the actor's partner is 'local' or 'external' (as in the model proposed by Guercini and Runfola 2015) is because firms not only absorb and disseminate within the same cluster (Camisón and Forés 2011), but can also absorb knowledge from outside the cluster and then disseminate it internally (Giuliani 2005).

Therefore, we believe that it is interesting to deep, in our research, the absorption and dissemination role analysing their degree (i.e. high or low), as proposed in Giuliani (2003). Hence, our research questions are:

RQ1: which is the degree of absorption and dissemination, with respect to I.4.0, of firms within ID?

RQ2: in such context, what is their role within those IDs?

Figure 1 represents our theoretical framework that summarizes four possible roles that each firm, within the ID, can adopt with respect to I.4.0: (i) 'traditionalists', (ii) 'bridges', (iii) 'isolated' and (iv) 'opportunists'.

'Bridges' and 'isolated firms' are at the two extremes. The firsts show high absorption and dissemination, acquiring technologies, competencies and behaviours 4.0 but, at the same time, being able to in turn disseminate them in the ID. The second ones are those who demonstrate poor I.4.0 attitudes and connections, since they do not assimilate significant technologies and attitudes 4.0 and therefore could not even spill them to the ID.

'Traditionalists' and 'opportunists' show opposite behaviours. On the one side, 'traditionalists' are not at the edge of innovation; however, they massively interact with other actors of the ID showing a high degree of dissemination. On the other side, 'opportunists' are those who take the most, benefiting from the advantages given by IDs when promoting I.4.0, but they do not give back to the same extent since they report low dissemination.

If it is true that absorptive capacity and dissemination are pivotal in case of disruptive innovation (Wang, Ling, and Chok 2020), when considering a revolutionary approach that radically changes the social and economic environment (as I.4.0) they seem to be the basis to understand the role that firms could play in an ID. Italian districts are mainly composed of small- and medium-sized manufacturing firms (Camuffo and

		Absorption	
		Low	High
Dissemination	High	Traditionalist	Bridge
	Low	Isolated	Opportunist

Figure 1. Theoretical framework.

Grandinetti 2011b), with an intricate plot of socio-economic relations (Becattini 1990). Therefore, it is important to understand the gatekeepers (Munari, Sobrero, and Malipiero 2012) and their role, and furthermore learn if, and how, some firms could be at the edge, having a boundary spanning role for the whole ID.

3. Methodology

3.1. Data collection

We interviewed 101 Italian firms of four Tuscan (Italy) IDs: the Santa Croce tanning district (16 firms), the Florentine leather district (42 firms), the Prato textile district (16 firms) and the Arezzo jewellery district (27 firms). The significance of those districts has been extensively studied by several scholars considering several aspects (Bacci, Lombardi, and Labory 2009; Becattini, Bellandi, and De Propris 2014; Bellandi and Dei Ottati 2016). More specifically, Lazzeretti (2003 and 2006), Rabino, Simoni, and Zanni (2008) and Zanni (2006) studied the importance of the Arezzo jewellery district for its economic impact, for the interesting managerial relationships of its firms and for being a significant case of the success of made in Italy. Prato textile district has been considered by Dei Ottati, Becattini, and Brusco (1995) – a paradigmatic case study for analysing and understanding Italian districts. Furthermore, it represents a significant example of integration of immigrant entrepreneurs in a unique socio-economic environment (Santini, Rabino, and Zanni 2011). The Florentine leather district features a significant number of network contracts (Milanesi, Guercini, and Tunisini 2020) taking advantage of the network instead of doing business by its own (Tunisini 2015). The Santa Croce tanning district has been considered by Ricciardi (2013) a meaningful example of cooperation among firms within a district, while Giannini and Valentino (2015) stressed the efficiency and the effectiveness of the district.

We used an ad hoc questionnaire prepared by scholars of four Tuscan Universities for the Tuscan Region (Fantoni et al. 2017a, 2017b) and used for assessing the level of I.4.0 maturity of manufacturing small and medium enterprises (SMEs). In Appendix A (see supplemental data) we reported the assessment used for interviewing the firms.

3.2. Measures

As anticipated, this paper looks at two dimensions: ‘absorption’ and ‘dissemination’. For the ‘absorption’, moving from Cohen and Levinthal (1990) and Nooteboom (2000), we recur to several information about competences, technologies 4.0, as well as culture 4.0, adapting what expressed in Bianchi and Bellini (1991), Bathelt (2001), Stuart and Sorenson (2003), Giuliani and Bell (2005), Majocchi and Presutti (2009) and Zhang and Li (2010).

Following the above-cited theory, we build the dimension 'absorption' considering three main areas: 'competencies 4.0', 'technologies 4.0' and 'culture 4.0'. Each one is composed by several sub-areas which, in turn, are obtained using the questionnaire that we reported in Appendix B (see supplemental data). More precisely, the area competencies 4.0 is composed by 'profiles 4.0', computer 'competencies' within the firm, and 'knowledge' of the firm's processes by the employees. The 'technologies 4.0' and 'culture 4.0' areas' sub-areas are reported in Appendix B.

For the dimension 'dissemination' we based our reasoning upon what emerged by the literature. Starting from the idea that some firms may act as gatekeepers (Morrison 2008), we adapted the concept written in Boschma and Ter Wal (2007), from a dissemination into the ID of 'non-local external knowledge' into a 'dissemination' of not-possessed knowledge, instead. Hence, to build the dimension, we consider, also considering the point of view of Santoni and Zanni (2012), the data collected about communication (both internal and external) (Ganesan, Malter, and Rindfleisch 2005; Novelli, Schmitz, and Spencer 2006), and collaborations (with several partners along the whole chain also and with research centres, universities, institutions, etc.) (Fromhold-Eisebith and Eisebith 2005; Novelli, Schmitz, and Spencer 2006).

As in the case of 'absorption', also for 'dissemination' we consider different areas for building the parameter. Therefore, we used 'communications' and 'collaborations' as the two main areas that are, in turn, composed of a portion of the assessment questionnaire we used for the interviews. For more details on how the two areas are composed, please read Appendix B. The two dimensions are used to understand how much I.4.0 is implemented within the firm (I.4.0 own level), i.e. 'absorption', and to understand how much it is then spilled towards the other firms of the supply chain, i.e. 'dissemination'. They have been measured on a scale from 1 to 6. Furthermore, to typify the actors' role, we also consider their 'integration'. This is because in local economic systems, firms are specialized in one or more core activities, thus often they could decide to outsource others to third parties. The more the firm outsources activities, the more it is connected with other actors. Hence, among other possible alternatives, we chose 'integration' for its important help in better characterising the actors' role. Therefore, we investigated which functions are performed directly and internally by the firm and which ones are made by partners. The more functions are exerted by external firms, the more the firm is eradicated in the supply chain. For 'integration' we adapted the concepts written in Mudambi and Venzin (2010) and Dibiaggio (2007), according to which outsourcing leads to a reduction of firm's personal integration (i.e. all or most of the functions are performed internally) and an increasing of collaborative integration (when some functions are outsourced).

We identified three degrees of 'integration': low, medium and high, based on the number of the functions externalized by the firm, considering all the possible functions that could be executed by a firm. As written before, integration represents a crucial aspect of I.4.0. Please read Appendix B for more information about integration.

4. Results

We decide to report the results of the firms belonging to the four IDs using four graphs, also adding another information we collected from the database Aida Bureau van Dijk: firms' dimensions (measured according to the guidelines of the European Commission 2020).

In the *X*-axis it is shown the 'absorption', while in the *Y*-axis we reported the 'dissemination' of the firms. The size of the spheres represents the level of 'integration'. Textures help in understanding the 'size': small squares if it is a small, vertical stripes if it is a medium, and dots if it is a large firm.

In Figure 2 are shown the 16 firms of the Santa Croce tanning district.

In Figure 3, we reported the 42 firms of the Florentine leather district, in Figure 4 are shown the 16 firms of the Prato textile industry, while in Figure 5 there are 27 firms of the Arezzo jewellery district.

5. Discussion

From the results, we immediately notice some differences between the four analysed IDs. For a more comprehensive comparison Figure 6 includes all the four districts. In oblique tripes the spheres of the Santa Croce tanning district; in vertical stripes the spheres of the

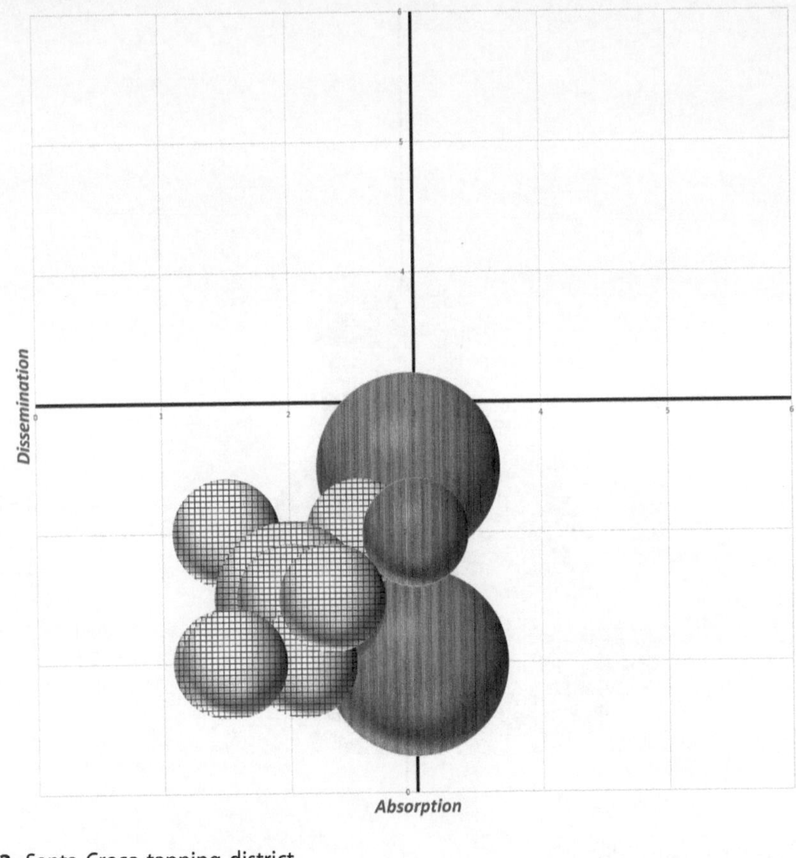

Figure 2. Santa Croce tanning district.

Prato textile district; in dots texture the spheres of the Florentine leather district; and in small squares' texture the spheres of the Arezzo jewellery district.

In the Santa Croce tanning district almost all the firms are in the bottom left quarter. This means that all the firms have both low 'absorption' and low 'dissemination'. From the size of the spheres, we also notice that those with medium/high integration are those in a better place, with only one firm being almost at the centre of the graph (3 'absorption' and 3 'dissemination').

If we move to the Florentine leather district, 'absorption' reaches higher levels, while most of the firms are still lacking in significant 'dissemination'. Here we start noticing that some firms occupy not only the bottom right quarter, but also the top left and top right one, meaning that some of them are in contrast with the average behaviour. A similar result is shown in the textile district of Prato. While 'absorption' shows close results to the Florentine leather district, with 'dissemination' it is possible to notice some better results from the firms in the textile district of Prato.

The most significant and quite surprising results come from the Arezzo jewellery district. Here a significant number of firms are in the top right quarter. There are still firms with poor 'absorption' and low 'dissemination', but we do also have excellence.

Considering the literature about IDs, we also would like to understand the role of firms according to their graph position. As written in the theoretical

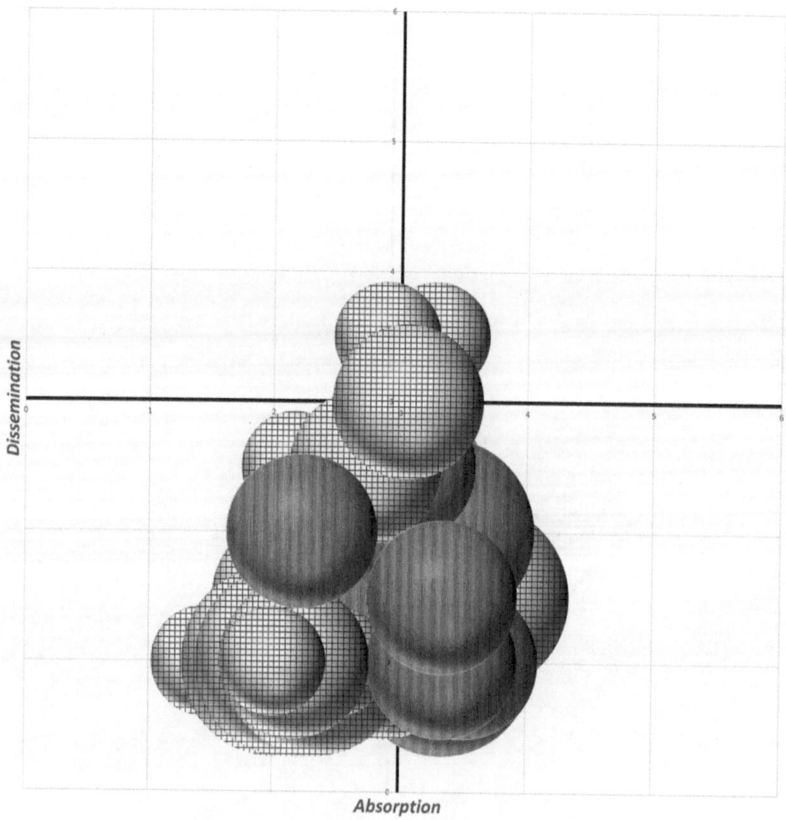

Figure 3. Florentine leather district.

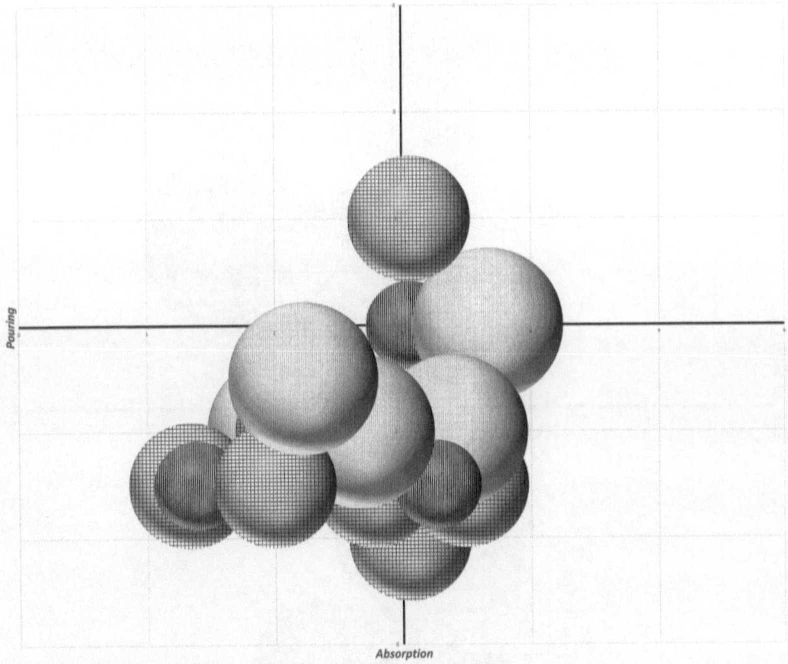

Figure 4. Prato textile industry.

framework's section, we have: (i) 'traditionalists', (ii) 'bridges', (iii) 'isolated' and (iv) 'opportunists'.

The two most interesting cases are, for opposite reasons, in the top right quarter and the bottom left ones. The 'isolated' behave as entities disconnected from the whole district. The best scenario comes from 'bridges', firms whose role is to adopt I.4.0 technologies and, more generally, of understanding and fully implementing the I.4.0 paradigm with the task to redistribute this knowledge by the dissemination activity towards the other firms of the ID. The firms with high dissemination and low absorption are very few. This could be due to the lack of dissemination of tacit knowledge, which should be the kind of knowledge disseminated by firms that absorb very little from outside, still having something (i.e. tacit knowledge) to disseminate. I.4.0 deals with codified knowledge, therefore tacit one is obsolete. I.4.0 eliminates the possibility to disseminate without absorbing, because without having the possibilities or capabilities or even the will to absorb I.4.0 (and therefore its codified knowledge) a firm cannot disseminate, since interconnections are strictly related to I.4.0 itself. Without this, it is not possible for firms to disseminate.

It is interesting to notice the firms' behaviour with the same size in different IDs: while in the tanning district the biggest firms (medium size) are the leading ones, also in terms of integration (sphere size), in the Florentine leather district the top performers are the small ones. All the medium size firms do not pour significantly towards the other players of the ID, while the only bridging firm is a small one. In the Jewellery district differences among firms' size are not so clear. It emerges that big firms are not as able to absorb and then pour technologies and ideas about I.4.0 as medium or small ones, medium firms with a medium degree of integration seem to be the most likely to exert a bridging role within the district. However, the inability to better play a bridging role might

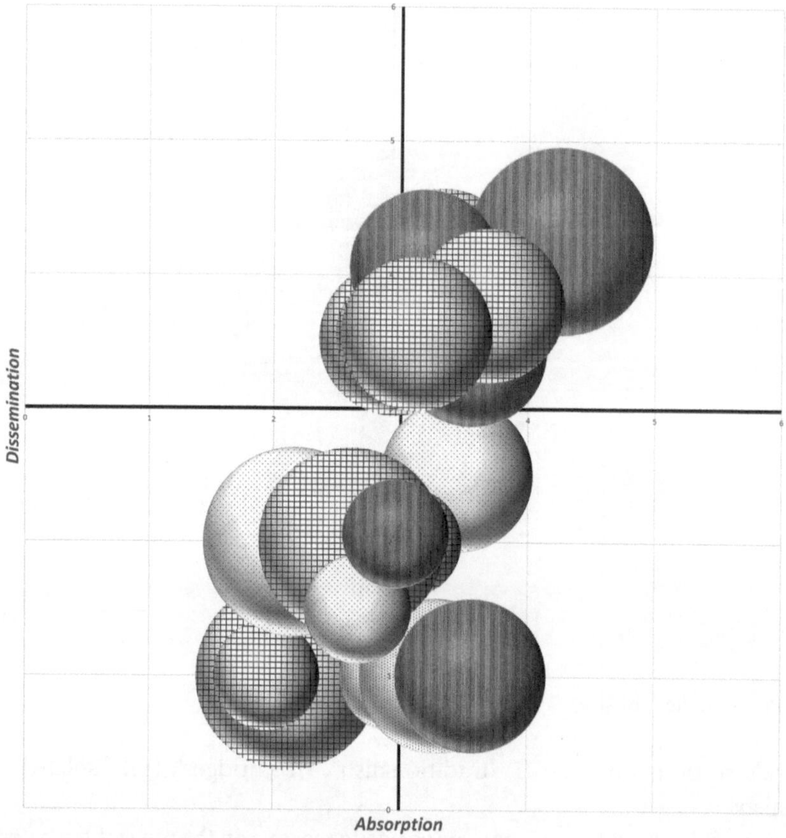

Figure 5. Arezzo jewellery district.

come from the lack from big firms to be flexible and agile in case of changes, while medium firms, with a sufficient degree of integration own, from one side, the resources and competences to assimilate I.4.0 and then diffuse it in the district.

If what discussed above seems reasonable, then it should appear strange the behaviour of the firms in the Florentine leather district. A possible explanation to this could come from a peculiarity of this industry. Here many firms are producing not for their own collection, using their own brand, but for bigger firms, mostly fashion multinationals. With no need to exploit internally own resources, since help comes directly from the third party, maybe a lighter and hence more flexible and dynamic entity could be able to capitalize on benefits more than a medium size firm.

6. Conclusion

Following what promised with the RQs, i.e. understanding degree of absorption and dissemination with respect to I.4.0 and firms' role, this study extends literature on IDs providing an in-depth analysis about how firms behave with respect to I.4.0 (De Propris and Bailey 2020). In so doing, this paper adds to extant research that has previously investigated absorption and dissemination (Giuliani 2003; Guercini and Runfola 2015) within IDs, providing a very nuanced picture of the heterogeneity of actors in four Tuscan IDs.

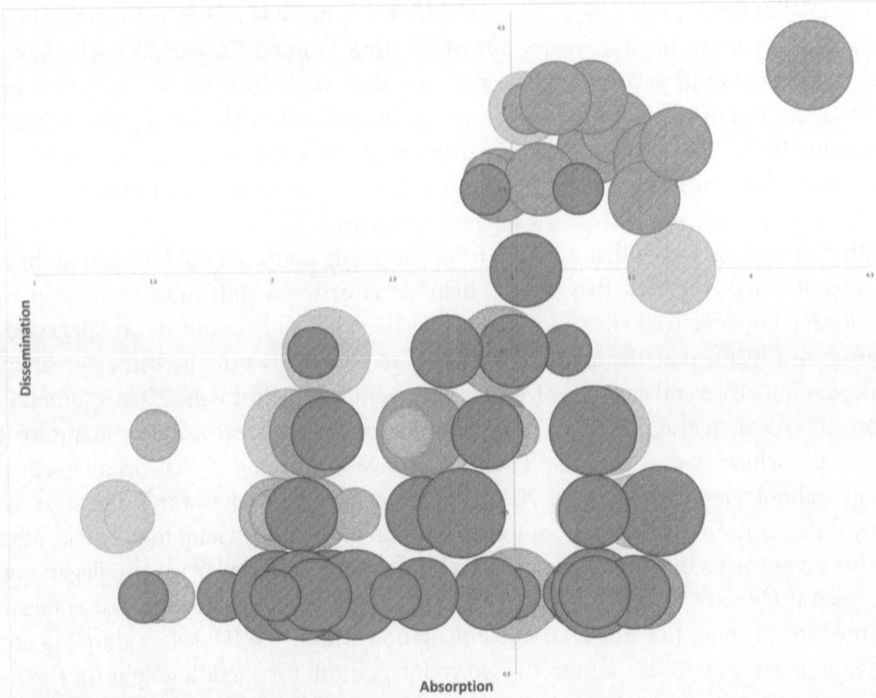

Figure 6. Comparison of all the four districts.

Although the number of districts is relatively small, some different behaviours amongst – and within them – emerged, confirming what also discussed considering different sectors in Büchi, Cugno, and Castagnoli (2020).

An interesting result coming from our research, that seems in contrast with what emerges from the analysis of previous works (Mittal et al. 2018; Horváth and Szabó 2019; Yu and Schweisfurth 2020), is the ability of small firms, in certain IDs, in the bridging activity. While the above-cited authors stressed the difficulties of SMEs in comparison to big firms, when trying to adopt I.4.0, we notice, in some IDs, a higher flexibility of smaller firms. This, together with the possibility to operate in a flexible and adjustable value chain, is in line with what stated by several authors, both referring to innovation in general and I.4.0 in particular. They point out the advantages of small, specialized firms within IDs with respect to bigger non-ID-firms or to firms in IDs that have an un-optimal big size in relation to the technical and flexibility optimum (Becattini 1990; Becattini, Bellandi, and De Propris 2014; Pegoraro, De Propris, and Chidlow 2020).

Furthermore, this could be due to the peculiar network that characterize the Florentine district (Tunisini 2015; Milanesi, Guercini, and Tunisini 2020) that could support and reinforce the absorptive capacity (De Propris and Bailey 2020) of small firms.

Albeit the number of IDs analysed as well as the number of firms within each IDs could represent a limitation, we think that this would be a good starting point for extending our research to other IDs from other Italian regions. Furthermore, we recommend a qualitative approach, as a polar multiple case study design, to understand why firms play a role rather than another.

This study has some implications for practitioners, since it shows in a direct way, using an immediate method the state of the art of the four studied Tuscan districts. Starting from this, firms should ask themselves the role they have in the IDs. 'Isolated' firms might lose the opportunity to benefit from an interaction with the district (Camuffo and Grandinetti 2011b), also because they display low 'integration'. Managers and entrepreneurs must be aware that in the long run this might represent a problem for them, especially if policy makers will act as suggested below.

Another important aspect that emerges from this study concerns the ID without 'bridge' firms, since it comes out that they should help some firms to shift towards the top right quarter for having beneficial effects for the whole ID. The study confirms, as suggested by Grashof et al. (2020), that the impact of I.4.0 onto the ID is not the same for anyone, hence depending on several factors which may vary significantly from one ID to another. Furthermore, this research also points out policy implications. If it is clear from the literature that the first stage (which started with the 'Piano Nazionale Industria 4.0') has been effective by financing technologies (Bettiol et al. 2020), now we have moved towards the next stage (stage two) that concerns the support of the integration of the value chain to fully take advantage of the investments in technologies, as also pointed out in a similar way by Pagano et al. (2020). Even if Hervas-Oliver, Di Maria, and Bettiol (2021) refer to actions, actors or strategies aimed for pushing towards I.4.0 implementation within the IDs when thinking about 'collective innovation polices', we use the word 'integration' for shedding light on the need for policies that incentivize a digital integration along the whole value chain of IDs. Recent policies hinged on I.4.0 applications and diffusion within single firms. For stage two we propose a collective implementation of I.4.0, then district-level focused policies, encouraging IDs to enforce the I.4.0 paradigm not at firm-level (within firms) but involving several IDs' firms (among firms) that will then be more integrated and will share common technologies and practices. This should be the task of the policy makers in the coming years. The proposed implication for policy makers is also in line with what also indicated by Büchi, Cugno, and Castagnoli (2020). They stressed the need for an intervention in several stages of the value chain, while we stress more the need for an intervention alongside the whole value chain that must be more punctual and with more precise effects than the ones adopted until now.

From this research, it emerges that neither firm's size nor technologies' adoption alone is crucial for an ID move towards 4.0. Competencies 4.0, culture 4.0 and collaboration (i.e. 'absorption' and 'dissemination') represent the pivotal mix for a District 4.0 move.

To the best of our knowledge, this research represents preliminary research on this issue from this specific point of view. Hence it improves the current knowledge of academics, policy makers and practitioners by proposing an analysis of the role of firms in Italian districts about I.4.0.

Acknowledgements

The assessment used for the 101 interviews is available upon request (Appendix A).

Disclosure statement

No potential conflict of interest was reported by the author(s).

Funding

The research has partially used the pre-assessment database from a research project co-financed by POR FESR 2014-2020 of Tuscany Region, Stazione Sperimentale per l'Industria delle Pelli e delle Materie Concianti, University of Siena, University of Florence and University of Pisa.

ORCID

Niccolò Fiorini http://orcid.org/0000-0002-8734-5858
Tommaso Pucci http://orcid.org/0000-0002-4404-9653
Elena Casprini http://orcid.org/0000-0001-5097-8793
Lorenzo Zanni http://orcid.org/0000-0002-0440-8842

References

Alcácer, J., J. Cantwell, and L. Piscitello. 2016. "Internationalization in the Information Age: A New Era for Places, Firms, and International Business Networks?". *Journal of International Business Studies* 47: 499–512. doi:10.1057/jibs.2016.22.

Almada-Lobo, F. 2015. "The Industry 4.0 Revolution and the Future of Manufacturing Execution Systems (MES)." *Journal of Innovation Management* 3 (4): 16–21. doi:10.24840/2183-0606_003.004_0003.

Álvarez Marcos, J., M. Capelo Hernández, and J. Álvarez Ortiz. 2019. "La madurez digital de la prensa española. Estudio de caso." *Revista Latina de Comunicación Social* 74: 499–520.

Bacci, L., M. Lombardi, and S. Labory. 2009. "The Evolution of External Linkages and Relational Density in the Tuscan Leather Industry." In *Business Networks in Clusters and Industrial Districts*, edited by Fiorenza Belussi and Alessia Sammarra, 176–201. London: Routledge.

Bathelt, H. 2001. "Regional Competence and Economic Recovery: Divergent Growth Paths in Boston's High Technology Economy." *Entrepreneurship and Regional Development* 13 (4): 287–314. doi:10.1080/08985620110067502.

Baur, C., and D. Wee. 2015. *Manufacturing's Next act.* Munich: McKinsey and Company.

Becattini, G. 1990. "The Marshallian Industrial District as a Socio-Economic Notion." Industrial Districts and Inter-Firm Co-Operation in Italy 2.

Becattini, G., M. Bellandi, and L. De Propris, eds. 2014. *A Handbook of Industrial Districts.* Cheltenham: Edward Elgar Pub.

Bellandi, M., and L. De Propris. 2017. "New Forms of Industrial Districts." *Economia e Politica Industriale* 44 (4): 411–427. doi:10.1007/s40812-017-0082-9.

Bellandi, M., L. De Propris, and E. Santini. 2019. "Industry 4.0 + Challenges to Local Productive Systems and Place Based Integrated Industrial Policies." In *Transforming Industrial Policy for the Digital age*, edited by Patrizio Bianchi, Clemente Ruiz Durán, and Sandrine Labory. Cheltenham: Edward Elgar Publishing.

Bellandi, M., and G. Dei Ottati. 2016. "Origine E Storia Dei Distretti Industriali Toscani." In *La Storia ILLUSTRATA Dell'economia Toscana Dall'ottocento Ad Oggi*, edited by Marco Bellandi, Stefano Casini Benvenuti, Marco Cini, Giuseppe Conti, Gabi Dei Ottati, Renato Giannetti, Monika Poettinger, and Piero Roggi, 105–150. Pisa: Pacini Editore.

Bellandi, M., E. Santini, C. Vecciolini, and L. De Propris. 2020. "Industry 4.0: Transforming Local Productive Systems in the Tuscany Region." In *Industry 4.0 and Regional Transformations*, edited by Lisa De Propris and David Bailey, 84–96. London: Routledge.

Belussi, F. 2005. "Are Industrial Districts Formed by Networks Without Technologies? The Diffusion of Internet Applications in Three Italian Clusters." *European Urban and Regional Studies* 12 (3): 247–268. doi:10.1177/0969776405056592.

Bettiol, M., M. Capestro, V. De Marchi, E. Di Maria, and S. Sedita. 2020. "Industrial Districts and the Fourth Industrial Revolution." *Competitiveness Review* 31 (1): 12–26. doi:10.1108/CR-12-2019-0155.

Bianchi, P., and N. Bellini. 1991. "Public Policies for Local Networks of Innovators." *Research Policy* 20 (5): 487–497. doi:10.1016/0048-7333(91)90071-W.

Bibby, L., and B. Dehe. 2018. "Defining and Assessing Industry 4.0 Maturity Levels–Case of the Defence Sector." *Production Planning and Control* 29 (12): 1030–1043. doi:10.1080/09537287.2018.1503355.

Bienhaus, F., and A. Haddud. 2018. "Procurement 4.0: Factors Influencing the Digitisation of Procurement and Supply Chains." *Business Process Management Journal* 24 (4): 965–984. doi:10.1108/BPMJ-06-2017-0139.

Boschma, R., and A. Ter Wal. 2007. "Knowledge Networks and Innovative Performance in an Industrial District: The Case of a Footwear District in the South of Italy." *Industry and Innovation* 14 (2): 177–199. doi:10.1080/13662710701253441.

Büchi, G., M. Cugno, and R. Castagnoli. 2020. "Smart Factory Performance and Industry 4.0." *Technological Forecasting and Social Change* 150: 119790. doi:10.1016/j.techfore.2019.119790.

Caloffi, A., L. Lazzeretti, and S. Sedita. 2018. "The Story of Cluster as a Cross-Boundary Concept: From Local Development to Management Studies." In *Agglomeration and Firm Performance*, edited by Fiorenza Belussi and Jose-Luis Hervas-Oliver, 123–137. Cham: Springer.

Camisón, C., and B. Forés. 2011. "Knowledge Creation and Absorptive Capacity: The Effect of Intra-District Shared Competences." *Scandinavian Journal of Management* 27 (1): 66–86. doi:10.1016/j.scaman.2010.11.006.

Camuffo, A., and R. Grandinetti. 2011a. "I distretti industriali come sistemi locali di innovazione." *Sinergie Italian Journal of Management* 69: 33–60.

Camuffo, A., and R. Grandinetti. 2011b. "Italian Industrial Districts as Cognitive Systems: Are They Still Reproducible?" *Entrepreneurship and Regional Development* 23 (9-10): 815–852. doi:10.1080/08985626.2011.577815.

Cohen, W., and D. Levinthal. 1990. "Absorptive Capacity: A New Perspective on Learning and Innovation." *Administrative Science Quarterly* 35 (1): 128–152. doi:10.2307/2393553.

Dalenogare, L., G. Benitez, N. Ayala, and A. Frank. 2018. "The Expected Contribution of Industry 4.0 Technologies for Industrial Performance." *International Journal of Production Economics* 204 (2018): 383–394. doi:10.1016/j.ijpe.2018.08.019.

De Propris, L., and D. Bailey. 2020. *Industry 4.0 and Regional Transformations*, 276. London: Taylor and Francis.

Dei Ottati, G., G. Becattini, and S. Brusco. 1995. *Tra mercato e comunità: aspetti concettuali e ricerche empiriche sul distretto industriale.* Milano: Franco Angeli.

Dibiaggio, L. 2007. "Design Complexity, Vertical Disintegration and Knowledge Organization in the Semiconductor Industry." *Industrial and Corporate Change* 16 (2): 239–267. doi:10.1093/icc/dtm006.

Ding, B. 2018. "Pharma Industry 4.0: Literature Review and Research Opportunities in Sustainable Pharmaceutical Supply Chains." *Process Safety and Environmental Protection* 119: 115–130. doi:10.1016/j.psep.2018.06.031.

Ericson, Å, J. Lugnet, J. Wenngren, H. Kaartinen, S. Pieskä, W. Solvang, and G. Sziebig. 2018. "Innovations and Industrial Internet: Research for Regional Growth and Competitiveness." 2018 2nd International Symposium on Small-Scale Intelligent Manufacturing Systems (SIMS), 1–4, IEEE.

European Commission. 2020. *User Guide to the SME Definition. European Commission, Brussels: Enterprise and Industry Publications.* Luxembourg: Publications Office of the European Union.

Fantoni, G., G. Cervelli, S. Pira, L. Trivelli, C. Mocenni, R. Zingone, and T. Pucci. 2017a. Ecosistemi 4.0: Imprese, Società, Capitale Umano, Fondazione G. Brodolini, Roma.

Fantoni, G., G. Cervelli, S. Pira, L. Trivelli, C. Mocenni, R. Zingone, and T. Pucci. 2017b. *Impresa 4.0: Siamo Pronti Alla Quarta Rivoluzione Industriale?* Pisa: Towel Publishing S.r.l.s, Pisa.

Fatorachian, H., and H. Kazemi. 2021. "Impact of Industry 4.0 on Supply Chain Performance". *Production Planning & Control*, 32 (1): 63–81.

Fromhold-Eisebith, M., and G. Eisebith. 2005. "How to Institutionalize Innovative Clusters? Comparing Explicit top-Down and Implicit Bottom-up Approaches." *Research Policy* 34 (8): 1250–1268. doi:10.1016/j.respol.2005.02.008.

Ganesan, S., A. Malter, and A. Rindfleisch. 2005. "Does Distance Still Matter? Geographic Proximity and New Product Development." *Journal of Marketing* 69 (4): 44–60. doi:10.1509/jmkg.2005.69.4.44.

Garzoni, A., I. De Turi, G. Secundo, and P. Del Vecchio. 2020. "Fostering Digital Transformation of SMEs: A Four Levels Approach." *Management Decision* 58 (8): 1543–1562. doi:10.1108/MD-07-2019-0939.

Ghadge, A., M. Kara, H. Moradlou, and M. Goswami. 2020. "The Impact of Industry 4.0 Implementation on Supply Chains." *Journal of Manufacturing Technology Management* 31 (4): 669–686. doi:10.1108/JMTM-10-2019-0368.

Ghobakhloo, M. 2018. "The Future of Manufacturing Industry: A Strategic Roadmap Toward Industry 4.0." *Journal of Manufacturing Technology Management* 29 (6): 910–936. doi:10.1108/JMTM-02-2018-0057.

Giannini, M., and T. Valentino. 2015. *Un modello di efficacia e di efficienza: Il distretto della pelle di Santa Croce sull'Arno*, 1–223. Rome: Aracne Editore.

Giuliani, E. 2003. "Cluster Absorptive Capacity: A Comparative Study Between Chile and Italy." Paper presentado en la conferencia a honor de keith pavitt "What Do We Know About Innovation".

Giuliani, E. 2005. "Cluster Absorptive Capacity: Why Do Some Clusters Forge Ahead and Others lag Behind?" *European Urban and Regional Studies* 12 (3): 269–288. doi:10.1177/0969776405056593.

Giuliani, E., and M. Bell. 2005. "The Micro-Determinants of Meso-Level Learning and Innovation: Evidence from a Chilean Wine Cluster." *Research Policy* 34 (1): 47–68. doi:10.1016/j.respol.2004.10.008.

Götz, M., and B. Jankowska. 2017. "Clusters and Industry 4.0–Do They Fit Together?" *European Planning Studies* 25 (9): 1633–1653. doi:10.1080/09654313.2017.1327037.

Grashof, N., A. Kopka, C. Wessendorf, and D. Fornahl. 2020. "Industry 4.0 and Clusters: Complementaries or Substitutes in Firm's Knowledge Creation?" *Competitiveness Review: An International Business Journal* 31 (1): 83–105. doi:10.1108/CR-12-2019-0162.

Guercini, S., and A. Runfola. 2015. "Actors' Roles in Interaction and Innovation in Local Systems: A Conceptual Taxonomy." *Journal of Business and Industrial Marketing* 30 (3/4): 269–278. doi:10.1108/JBIM-12-2012-0256.

Halldórsson, Á, J. Hsuan, and H. Kotzab. 2015. "Complementary Theories to Supply Chain Management Revisited-from Borrowing Theories to Theorizing." *Supply Chain Management: An International Journal* 20 (6): 574–586. doi:10.1108/SCM-06-2015-0228.

Hervas-Oliver, J., E. Di Maria, and M. Bettiol. 2021. "Industry 4.0 in Firms, Clusters and Regions: The New Digital Imperative." *Competitiveness Review: An International Business Journal* 31 (1): 1–11. doi:10.1108/CR-01-2021-164.

Hervas-Oliver, J., S. Estelles-Miguel, G. Mallol-Gasch, and J. Boix-Palomero. 2019. "A Place-Based Policy for Promoting Industry 4.0: The Case of the Castellon Ceramic Tile District." *European Planning Studies* 27 (9): 1838–1856. doi:10.1080/09654313.2019.1642855.

Hervás-Oliver, J. 2021. "Industry 4.0 in Industrial Districts: Regional Innovation Policy for the Toy Valley District in Spain." *Regional Studies* 55 (10-11): 1775–1786. doi:10.1080/00343404.2021.1939861.

Horváth, D., and R. Szabó. 2019. "Driving Forces and Barriers of Industry 4.0: Do Multinational and Small and Medium-Sized Companies Have Equal Opportunities?" *Technological Forecasting and Social Change* 146: 119–132. doi:10.1016/j.techfore.2019.05.021.

Ilvonen, I., S. Thalmann, M. Manhart, and C. Sillaber. 2018. "Reconciling Digital Transformation and Knowledge Protection: A Research Agenda." *Knowledge Management Research and Practice* 16 (2): 235–244. doi:10.1080/14778238.2018.1445427.

Innocenti, N., F. Capone, and L. Lazzeretti. 2020. "Knowledge Networks and Industrial Structure for Regional Innovation: An Analysis of Patents Collaborations in Italy." *Papers in Regional Science* 99 (1): 55–72. doi:10.1111/pirs.12478.

Kagermann, H. 2015. "Change Through Digitization—Value Creation in the age of Industry 4.0." In *Management of Permanent Change*, edited by Horst Albach, Heribert Meffert, Andreas Pinkwart, and Ralf Reichwald, 23–45. Wiesbaden: Springer Gabler.

Kagermann, H., W. Wahlster, and J. Helbig. 2013. "Securing the Future of German Manufacturing Industry: Recommendations for Implementing the Strategic Initiative INDUSTRIE 4.0." Final Report of the Industrie 4.

Kaltenbach, F., P. Marber, C. Gosemann, T. Bölts, and A. Kühn. 2018. "Smart Services Maturity Level in Germany." 2018 IEEE International Conference on Engineering, Technology and Innovation (ICE/ITMC), 1–7, IEEE.

Kipper, L., L. Furstenau, D. Hoppe, R. Frozza, and S. Iepsen. 2020. "Scopus Scientific Mapping Production in Industry 4.0 (2011-2018): A Bibliometric Analysis." *International Journal of Production Research* 58 (6): 1605–1627. doi:10.1080/00207543.2019.1671625.

Kohtamäki, M., V. Parida, P. Oghazi, H. Gebauer, and T. Baines. 2019. "Digital Servitization Business Models in Ecosystems: A Theory of the Firm." *Journal of Business Research* 104 (11): 380–392. doi:10.1016/j.jbusres.2019.06.027.

Koren, Y., and M. Shpitalni. 2010. "Design of Reconfigurable Manufacturing Systems." *Journal of Manufacturing Systems* 29 (4): 130–141. doi:10.1016/j.jmsy.2011.01.001.

Kulyk, V., and D. Škodová Parmová. 2017. "E-business Development: The Comparative Study of the Czech Republic and the Ukraine." *Deturope* 9: 80–110. doi:10.32725/det.2017.007.

Lasi, H., P. Fettke, H. Kemper, T. Feld, and M. Hoffmann. 2014. "Industry 4.0." *Business and Information Systems Engineering* 6 (4): 239–242. doi:10.1007/s12599-014-0334-4.

Lazzeretti, L. 2003. *Nascita ed evoluzione del distretto orafo di Arezzo, 1947-2001*. Florence: Firenze University Press.

Lazzeretti, L. 2006. "Density Dependent Dynamics in the Arezzo Jewellery District (1947–2001): Focus on Foundings." *European Planning Studies* 14 (4): 431–458. doi:10.1080/09654310500421055.

Lazzeretti, L., and F. Capone. 2009. *Industrial District Effects and Innovation in the Tuscan Shipbuilding Industry. IERMB Working Paper in Economics 0903*. Barcelona: Institut d'Estudis Regionals i Metropolitans de Barcelona.

Lazzeretti, L., and F. Capone. 2015. "Innovations and Innovators in a Resilient City: The Case of Chemical Innovations After the 1966 Flood in Florence City." *Culture and Society* 6 (3): 83–91. doi:10.1016/j.ccs.2015.02.004.

Lazzeretti, L., and F. Capone. 2016. "How Proximity Matters in Innovation Networks Dynamics Along the Cluster Evolution. A Study of the High Technology Applied to Cultural Goods." *Journal of Business Research* 69 (12): 5855–5865. doi:10.1016/j.jbusres.2016.04.068.

Lazzeretti, L., and F. Capone. 2017. "Liabilities in Prato's Industrial District: An Analysis of Italian and Chinese Firm Failures." In *Native and Immigrant Entrepreneurship*, edited by Simone Guercini, Gabi Dei Ottati, Loretta Baldassar, and Graeme Johanson, 149–167. Cham: Springer.

Li, L. 2018. "China's Manufacturing Locus in 2025: With a Comparison of "Made-in-China 2025" and "Industry 4.0"." *Technological Forecasting and Social Change* 135: 66–74. doi:10.1016/j.techfore.2017.05.028.

Majeed, A.., and T. Rupasinghe. 2017. "Internet of Things (IoT) Embedded Future Supply Chains for Industry 4.0: An Assessment from an ERP-Based Fashion Apparel and Footwear Industry." *International Journal of Supply Chain Management* 6 (1): 25–40.

Majocchi, A., and M. Presutti. 2009. "Industrial Clusters, Entrepreneurial Culture and the Social Environment: The Effects on FDI Distribution." *International Business Review* 18 (1): 76–88. doi:10.1016/j.ibusrev.2008.12.001.

Malipiero, A., F. Munari, and M. Sobrero. 2005. "Focal Firms as Technological Gatekeepers Within Industrial Districts: Knowledge Creation and Dissemination in the Italian Packaging Machinery Industry." Danish Research Unit for Industrial Dynamics.

Milanesi, M., S. Guercini, and A. Tunisini. 2020. "Exploring SMEs' Qualitative Growth and Networking Through Formalization." *Competitiveness Review: An International Business Journal* 30 (4): 397–415.

Mittal, S., M. Khan, D. Romero, and T. Wuest. 2018. "A Critical Review of Smart Manufacturing and Industry 4.0 Maturity Models: Implications for Small and Medium-Sized Enterprises (SMEs)." *Journal of Manufacturing Systems* 49: 194–214. doi:10.1016/j.jmsy.2018.10.005.

Molina Morales, F., V. De Marchi, and L. Martínez-Cháfer. 2021. "Absorptive Capacity and Radical Innovation in Industrial Districts." *Technology Analysis and Strategic Management* 33 (9): 1088–1100. doi:10.1080/09537325.2020.1869207.

Molina-Morales, F., L. Martínez-Cháfer, and D. Valiente-Bordanova. 2017. "Disruptive Technological Innovations as New Opportunities for Mature Industrial Clusters. The Case of Digital Printing Innovation in the Spanish Ceramic Tile Cluster." *Investigaciones Regionales-Journal of Regional Research* 39: 39–57. http://hdl.handle.net/10017/31599.

Morrison, A. 2008. "Gatekeepers of Knowledge Within Industrial Districts: Who They are, How They Interact." *Regional Studies* 42 (6): 817–835. doi:10.1080/00343400701654178.

Mudambi, R., and M. Venzin. 2010. "The Strategic Nexus of Offshoring and Outsourcing Decisions." *Journal of Management Studies* 47 (8): 1510–1533. doi:10.1111/j.1467-6486.2010.00947.x.

Munari, F., M. Sobrero, and A. Malipiero. 2012. "Absorptive Capacity and Localized Spillovers: Focal Firms as Technological Gatekeepers in Industrial Districts." *Industrial and Corporate Change* 21 (2): 429–462. doi:10.1093/icc/dtr053.

Müller, J., O. Buliga, and K. Voigt. 2018. "Fortune Favors the Prepared: How SMEs Approach Business Model Innovations in Industry 4.0." *Technological Forecasting and Social Change* 132: 2–17. doi:10.1016/j.techfore.2017.12.019.

Müller, J., O. Buliga, and K. Voigt. 2021. "The Role of Absorptive Capacity and Innovation Strategy in the Design of Industry 4.0 Business Models – A Comparison Between SMEs and Large Enterprises." *European Management Journal* 39 (3): 333–343. doi:10.1016/j.emj.2020.01.002.

Nooteboom, B. 2000. "Learning by Interaction: Absorptive Capacity, Cognitive Distance and Governance." *Journal of Management and Governance* 4 (1): 69–92. doi:10.1023/A:1009941416749.

Novelli, M., B. Schmitz, and T. Spencer. 2006. "Networks, Clusters and Innovation in Tourism: A UK Experience." *Tourism Management* 27 (6): 1141–1152. doi:10.1016/j.tourman.2005.11.011.

Owen-Smith, J., and W. Powell. 2004. "Knowledge Networks as Channels and Conduits: The Effects of Spillovers in the Boston Biotechnology Community." *Organization Science* 15 (1): 5–21. doi:10.1287/orsc.1030.0054.

Pagano, A., E. Carloni, S. Galvani, and R. Bocconcelli. 2020. "The Dissemination Mechanisms of Industry 4.0 Knowledge in Traditional Industrial Districts: Evidence from Italy." *Competitiveness Review* 31 (1): 27–53. doi:10.1108/CR-12-2019-0160.

Parra-Requena, G., M. Ruiz-Ortega, and P. Garcia-Villaverde. 2013. "Social Capital and Effective Innovation in Industrial Districts: Dual Effect of Absorptive Capacity." *Industry and Innovation* 20 (2): 157–179. doi:10.1080/13662716.2013.771486.

Pegoraro, D., L. De Propris, and A. Chidlow. 2020. "De-Globalisation Value Chains and Reshoring". In *Industry 4.0 and Regional Transformations*, 152–175 edited by De Propris and Bailey. London: Taylor and Francis.

Piccarozzi, M., B. Aquilani, and C. Gatti. 2018. "Industry 4.0 in Management Studies: A Systematic Literature Review." *Sustainability* 10 (10): 3821. doi:10.3390/su10103821.

Pirola, F., C. Cimini, and R. Pinto. 2020. "Digital Readiness Assessment of Italian SMEs: A Case-Study Research." *Journal of Manufacturing Technology Management* 31 (5): 1045–1083. doi:10.1108/JMTM-09-2018-0305.

Pucci, T., L. Zanni, A. Runfola, and S. Guercini. 2018. "The Role of the Actors in Interactions Between "Innovation Ecosystems": Drivers and Implications." *IMP Journal (Merged with Journal of Business and Industrial Marketing)* 12 (2): 333–345. doi:10.1108/IMP-05-2017-0022.

Rabino, S., C. Simoni, and L. Zanni. 2008. "Small and Medium Gold and Fashion Enterprises (SMEs) in Arezzo, Italy: Competitive Challenges and Strategic Implications." *Journal of Global Marketing* 21 (2): 141–159. doi:10.1080/08911760802135525.

Ricciardi, A. 2013. "I distretti industriali italiani: recenti tendenze evolutive (Italian Industrial Districts: Recent Evolutionary Trends)." *Sinergie Italian Journal of Management* 31 (May-Aug): 21–58. doi:10.7433/s91.2013.03.

Rojko, A. 2017. "Industry 4.0 Concept: Background and Overview." *International Journal of Interactive Mobile Technologies* 11 (5): 77–90. doi:10.3991/ijim.v11i5.7072.

Rüßmann, M., M. Lorenz, P. Gerbert, M. Waldner, J. Justus, P. Engel, and M. Harnisch. 2015. "Industry 4.0: The Future of Productivity and Growth in Manufacturing Industries." *Boston Consulting Group* 9 (1): 54–89.

Rüttimann, B., and M. Stöckli. 2016. "Lean and Industry 4.0-Twins, Partners, or Contenders? A Due Clarification Regarding the Supposed Clash of Two Production Systems." *Journal of Service Science and Management* 9 (6): 485–500. doi:10.4236/jssm.2016.96051.

Santini, C., S. Rabino, and L. Zanni. 2011. "Chinese Immigrants Socio-Economic Enclave in an Italian Industrial District: The Case of Prato." *World Review of Entrepreneurship, Management and Sustainable Development* 7 (1): 30–51. doi:10.1504/WREMSD.2011.038312.

Santoni, S., and L. Zanni. 2011. "Comportamenti di ricerca e innovazione nei distretti industriali: gli attori del cambiamento e l'accesso a "conoscenze distanti." *Sinergie* 84: 231–256.

Schmidt, R., M. Möhring, R. Härting, C. Reichstein, P. Neumaier, and P. Jozinović. 2015. "Industry 4.0-Potentials for Creating Smart Products: Empirical Research Results." International Conference on Business Information Systems, June, 16–27. Cham: Springer.

Schuh, G., T. Potente, R. Varandani, and T. Schmitz. 2014. "Global Footprint Design Based on Genetic Algorithms-An "Industry 4.0" Perspective." *CIRP Annals* 63 (1): 433–436. doi:10.1016/j.cirp.2014.03.121.

Schumacher, A., S. Erol, and W. Sihn. 2016. "A Maturity Model for Assessing Industry 4.0 Readiness and Maturity of Manufacturing Enterprises." *Procedia Cirp* 52: 161–166. doi:10.1016/j.procir.2016.07.040.

Stuart, T., and O. Sorenson. 2003. "The Geography of Opportunity: Spatial Heterogeneity in Founding Rates and the Performance of Biotechnology Firms." *Research Policy* 32 (2): 229–253. doi:10.1016/S0048-7333(02)00098-7.

Suzuki, T., T. Noda, H. Shibuya, and I. Suzuki. 2014. "An Anomaly Detection System for Advanced Maintenance Services." *Hitachi Review* 63 (4): 178–182.

Takakuwa, S., I. Veza, and S. Celar. 2018. "INDUSTRY 4.0" IN EUROPE AND EAST ASIA." *Annals of DAAAM and Proceedings* 29: 61–69. doi:10.2507/29th.daaam.proceedings.009.

Thames, L., and D. Schaefer. 2016. "Software-defined Cloud Manufacturing for Industry 4.0." *Procedia Cirp* 52: 12–17. doi:10.1016/j.procir.2016.07.041.

Tunisini, A. 2015. "Percorsi di nascita e di sviluppo dei contratti di rete." *Econoomia e Società Regionale* 2015 (2): 60–75. doi:10.3280/ES2015-002005.

Vasin, S., E. Gamidullaeva, I. Shkarupeta, and T. Palatkin. 2018. "Emerging Trends and Opportunities for Industry 4.0 Development in Russia." *European Research Studies Journal* 21 (3): 63–76. doi:10.35808/ersj/1044.

Wang, Z., J. Ling, and J. Chok. 2020. "Relational Embeddedness and Disruptive Innovations: The Mediating Role of Absorptive Capacity." *Journal of Engineering and Technology Management* 57: 101587. doi:10.1016/j.jengtecman.2020.101587.

Winkelhaus, S., and E. Grosse. 2020. "Logistics 4.0: A Systematic Review Towards a new Logistics System." *International Journal of Production Research* 58 (1): 18–43. doi:10.1080/00207543.2019.1612964.

Witkowski, K. 2017. "Internet of Things, Big Data, Industry 4.0-Innovative Solutions in Logistics and Supply Chains Management." *Procedia Engineering* 182: 763–769. doi:10.1016/j.proeng.2017.03.197.

Xu, M., J. David, and S. Kim. 2018. "The Fourth Industrial Revolution: Opportunities and Challenges." *International Journal of Financial Research* 9 (2): 90–95. doi:10.5430/ijfr.v9n2p90.

Yu, F., and T. Schweisfurth. 2020. "Industry 4.0 Technology Implementation in SMEs-A Survey in the Danish-German Border Region." *International Journal of Innovation Studies* 4 (3): 76–84. doi:10.1016/j.ijis.2020.05.001.

Zanni, L., ed. 2006. *Distretti industriali e nuovi scenari competitivi: l'esperienza del distretto orafo aretino*. Milan: Angeli.

Zhang, Y., and H. Li. 2010. "Innovation Search of new Ventures in a Technology Cluster: The Role of Ties with Service Intermediaries." *Strategic Management Journal* 31 (1): 88–109. doi:10.1002/smj.806.

Robots and the productivity of local manufacturing systems in Emilia-Romagna: the mediating role of occupational similarity and complexity

Roberto Antonietti, Luca Cattani and Giulio Pedrini

ABSTRACT

During the last decade, a growing literature has been assessing the impact of industrial robots on productivity and employment. At the local level, the labour-displacing effect of robot adoption can be counterbalanced by a task reallocation between manufacturing industries characterized by similar/related occupational spaces. This reallocation process can be also influenced by the average degree of occupational complexity of the local labour markets. To test for the mediating role of similarity and complexity in the robot-employment-productivity relationship, we develop a two-step analysis mixing information from different data sources on 15 industries and 39 local labour market areas of the Emilia-Romagna region in Italy, for the period 2008–2017. Our analysis shows that higher exposure to robots does not affect the aggregate level of labour productivity, but, rather, is associated with a decline in (high routine) employment after three years. We further show that the negative relationship between robots and employment decreases, or even vanishes, in local labour markets with high cross-industry occupational similarity and low occupational or task complexity, and within industrial districts.

1. Introduction

The perception that robots 'will steal our jobs' is still high among people and policy-makers (McKinsey Global Institute 2017; World Economic Forum 2020). A recent Morning Consult Poll[1] found that almost half of the interviewed American citizens in 2021 fear that automation will reduce the number of jobs, this feeling being higher among adults earning less than $50,000 yearly, among those without a college degree, or those voting for the Republican party. For Italy, the country profile of the World Economic Forum (2020) shows that the emerging jobs are those concerning artificial intelligence and machine learning specialists, Internet of Things specialists, data analysts, scientists, digital transformation experts, or project managers. In other terms, the emerging jobs will be those where creativity, originality, initiative, analytic and critical thinking, innovation, and emotional intelligence will be the core skills. On the contrary, data entry clerks, administrative staff, bookkeepers, assemblers, factory workers, and human resource managers are seen as declining jobs because they will be replaced by automated machinery. As Acemoglu and Restrepo (2020) show, however, the impact of robots on employment is multifaceted and varies across regions and industries. In this paper, we argue that two elements of regions and industries that should be considered are the degree of cross-industry occupational similarity and the degree of occupational complexity.

With respect to the former, in the last two decades, studies in regional economics and economic geography developed the notions of related variety, and relatedness (Hidalgo et al. 2018), as a potential source of agglomeration externalities (Frenken, van Oort, and Verburg 2007). This notion has been initially applied to the industrial composition of local economies referring to sectors belonging to the same product family (e.g. Boschma, Eriksson, and Lindgren 2008) or patent class (Rigby 2015). More recently, the concept has evolved and referred to the educational background and occupation of employees (Neffke and Henning 2013; Muneepeerakul et al. 2013), making it closer to the one of occupational similarity (e.g. Neffke, Nedelkoska, and Wiederhold 2022). When applied to occupations, in fact, both relatedness and similarity can be used to measure the diffusion of occupations that are close to each other in terms of tasks and skills, the former being indicated for intra-industry linkages while the latter is better suited for inter-industry nexus (Nedelkoska and Neffke 2019; Wixe and Andersson 2017). These notions, and the related metrics, go beyond industrial and technological dimensions and are expected to catch both matching processes in local labour markets and knowledge spillovers, two major sources of agglomeration economies. Regions with diversified, but similar, jobs (as defined by their constituent tasks and skills) may ease the matching between employers and employees in the labour market, as well as inter-firm knowledge transfers through employee mobility (Wixe and Andersson 2017).

A similar path has been followed by the notion of economic complexity (Hidalgo and Hausmann 2009), which initially reflected the composition of an economy's productive structure and its position in the global trade network. As for relatedness, this notion of complexity has evolved and was recently applied not only to patent data and technology classes (Balland and Rigby 2017) but also to the occupational structure of countries and regions, using newly available information on tasks and skills (Lo Turco and Maggioni 2020; Antonietti et al. 2022). Countries or regions are considered more complex if they are endowed with a wider portfolio and a higher number of exclusive jobs/tasks, which cannot be easily copied and replicated elsewhere while they may complement and support the tasks accomplished by automated technology.

In the manufacturing industry, a major source of labour demand changes is robotization, a form of capital-embodied technological progress that is deemed to raise productivity by either substituting or complementing firms' workforce (Acemoglu and Restrepo 2020; Klenert, Fernandez-Macias, and Antón 2020). So far, the extant empirical evidence has confirmed that the use of industrial robots significantly increased both value-added and labour productivity (e.g. Gratz and Micheals 2018; Dauth et al. 2021; Kromann et al. 2020). The net effect of robot adoption on employment has also been thoroughly investigated (see Dottori 2021 for a review), but its sign and magnitude change over time and space as it encompasses contrasting forces going in opposite directions (Acemoglu and Restrepo 2018). What is missing in the literature is an analysis of the characteristics of local production systems, that, in principle, can mitigate the negative impact of robots on employment or can even favour the creation of new job profiles and new skills.

According to these premises, this paper aims to address the following research questions. Does a higher exposure to industrial robots increase the productivity of local manufacturing systems? If so, is this relationship driven by the (positive) impact of robots on value-added or by the (negative) impact on employment? Does the relationship between robots and employment also depend on the level of cross-industry occupational similarity and complexity of the local labour market? Finally, does this relationship change within and outside industrial districts?

To answer these questions, we combine information from four data sources. Data on robot exposure come from the World Robotics 2020 managed by the International Federation of Robotics. Data on employment and occupational titles come from a rich longitudinal employer-employee dataset available for the manufacturing activities of Emilia-Romagna, Italy, that we combine with a national survey provided by INAPP-ISTAT on the tasks and skills used on the job. Finally, we add balance sheet information coming from AIDA Bureau van Dijk and we obtain a sample of 39 local labour market areas and 15 two-digit manufacturing sectors over 10 years (2008–2017). Although focusing on one region restricts the scope of the analysis, Emilia-Romagna represents one of the main manufacturing regions in Europe, ranking among the top regions in 2020 for high employment and low unemployment rates.[2]

The paper is structured as follows. Section 2 provides the theoretical background for our first-step analysis concerning the relationship between robots, productivity, and employment. Section 3 describes the first empirical analysis, by introducing the data sources, the main variables, the econometric strategy, and illustrating the first set of results. Taking stock of these results, Section 4 provides the theoretical background for the second-step analysis on the mediating role of occupational similarity and complexity, as well as the possible role of industrial districts. Section 5 describes the second empirical analysis and the related second-step regression results. Finally, in Section 6 we take stock of the econometric analysis, and we discuss policy implications.

2. Robots, productivity and employment: related literature and theoretical framework

The massive diffusion of industrial robots in manufacturing production processes started more than forty years ago. Over time, their diffusion substantially increased in all western countries while their labour-replacing capabilities gradually encompassed a wider and wider range of human tasks (Brynjolfsson and McAfee 2012). With their increasing use in manufacturing processes, industrial robots have deeply affected both labour demand (Frey and Osborne 2017; Acemoglu and Restrepo 2020; Dauth et al. 2021; Bessen 2019) and labour productivity (e.g. Carbonero, Ernst, and Weber 2020; Graetz and Michaels 2018; Kromann et al. 2020). In the following Section, we review the literature that studied the links between robot adoption and productivity and robot adoption and employment, respectively.

2.1. Robots and productivity

The extant empirical literature is quite homogeneous when looking at the impact of robots on productivity. Most studies showed that robot density is a driver of labour productivity in manufacturing industries. Cross-country analyses show that the adoption of industrial robots increase labour productivity by 0.36% further to an increase of 1% in robot density (expressed as the number of robots per million hours worked), on average, between 1993 and 2007 (Graetz and Michaels 2018) while a higher impact (6.6%) concerns total factor productivity of nine industrialized countries for the period 2004–2007 (Kromann et al. 2020). Single-country analyses confirmed these positive effects for Canada, Germany, France, and Spain and for more recent years. Dixon et al. (2021) report positive effects of robots on firm productivity in the Canadian economy over the years 2000–2015. Dauth et al. (2021) show that exposure to robots induces a sizable and statistically significant increase in labour productivity in German local labour markets between 2004 and 2014. Acemoglu, Lelarge, and Restrepo (2020) find similar results for French manufacturing firms during the period 2010–2015 while showing that robot adopters also expand the number of hours worked, offsetting the job losses suffered by other firms. This result is confirmed and more thoroughly discussed by Koch, Manuylov, and Smolka (2021), who identify two sources of productivity gains stemming from robot adoption in Spanish manufacturing firms from 1990 to 2016. The first is the direct benefit associated with the use of robots in the production process. Such benefit can be task-neutral, resulting in uniform productivity increases in all tasks, but more often it is expected to be task-biased, as robot adoption is usually assumed to increase productivity through a reallocation of routine tasks away from workers toward capital (Acemoglu and Restrepo 2021) with the consequent substitution of high-routine jobs with robots. This hypothesis is in line with the routine-biased technical change approach (Autor, Levy, and Murnane 2003), which posits the job destruction effect of new technologies among substitutable occupations. The second source is a productivity-enhancing labour reallocation across firms, with non-negative employment effects in robot adopters, who were more likely to create new jobs in occupations characterized by a relatively high level of technical skills,[3] whereas firms that did not adopt robots tended to destroy jobs further to the loss of market shares. This effect entails the existence of complementarity between robots and occupational space at the industry level, leaving open the question of which geographical factors that may enable this productivity-enhancing worker reallocation within local labour markets to avoid job losses (Acemoglu and Restrepo 2020).

2.2. Robots and employment

The productivity advancements achieved by more intense use of robots might depend on the effect that this latter has on employment. From a theoretical standpoint, however, several forces are at play in shaping employment changes due to automation. First, as humans and computers compete for the allocation of routine task inputs, substitution/displacement is expected to take place as the latter become relatively cheaper and/or more productive (Autor, Levy, and Murnane 2003; Goos, Manning, and Salomons 2014; Acemoglu and Restrepo 2018). Robots' capability to perform an ever-increasing number of routines raised the concern about whether they will replace jobs in the future (Gentili et al. 2020). Second, automated tasks may become more productive by either reducing their relative costs or augmenting the relative outputs thus leading to a 'deepening of automation' or 'automation at the intensive margin' (Acemoglu and Restrepo 2018). However, this increased productivity brings about additional consequences that must be taken into account when trying to disentangle different countervailing shifts that can affect the overall occupational change driven by automation. In fact, the third force at play in this framework is represented by the productivity effect itself: declining prices of goods and services in one sector might expand its relative market, increasing its relative labour demand, as in the case of ATMs in the banking sector (Bessen 2019). Alternatively, big productivity pushes in one sector may have a sizable impact on the labour demand in other sectors too, as is the case for productivity gains in the agricultural sector that led to increased household purchasing power. When the increases in labour demand derived from productivity effects outpace the relative losses determined by the 'displacement effect', we have the so-called 'reinstatement effect' (Acemoglu and Restrepo 2019) entailing that the range of tasks that can be automated is expanded thus displacing human labour in a wider set of activities ('automation at the extensive margin').

Unsurprisingly, given the potential combination of contrasting effects, the recent empirical literature on this topic provides contrasting results Dottori (2021). The employment effect of robotization is not clear-cut and both a reduction and an increase in economy-wide labour demand can arise (De Vries et al. 2020). Like many new technologies, robots may replace jobs if one keeps output constant, but they may also stimulate labour demand expressed by other industries thanks to the increase the consumer demand entailed by lower prices or improved product quality (Dauth et al. 2021). This quantitative effect on labour demand is deeply intertwined with the impact of robots on the qualitative composition of labour demand (Dixon et al. 2021), which are deemed to follow education-based, task-biased, or complexity-biased patterns (Antonietti et al. 2022).

3. Empirical analysis: step 1

3.1. Data and variables

Our empirical analysis focuses on the manufacturing sectors in local labour market areas (LLMAs) of the Emilia-Romagna region in Italy. This is one of the most important European manufacturing regions, ranking fifth across all the EU for total employment. As in the rest of the country, micro, small and medium enterprises employ a relatively large share of the labour force compared to other OECD countries (OECD 2019), and tend to specialize in a set of specific industries according to their relative geographical location, combining both high and low market producers (Piore and Sabel 1985; Brusco 1982). Due to these characteristics, Emilia-Romagna offers an ideal mix of industry similarity, complexity, and heterogeneity to analyse the robot-employment-productivity nexus at the local level.

Our empirical analysis develops in two steps. First, we test whether higher exposure of LLMAs to robots corresponds to higher levels of labour productivity and whether such a relationship passes through a possible decrease in employment rather than through an increase in value-added. To do so, we combine four different data sources. Employment data come from the 'Sistema Informativo Lavoro – Emilia-Romagna' (SILER), which is an administrative Linked Employers-Employee microData (LEED) resulting from compulsory declarations employers address to the regional employment office for all occurrences concerning a job position (hiring, firing, conversion, and fixed-term contract prolonged duration). SILER provides information on the universe of manufacturing firms, and the corresponding labour force, between 2008 and 2017. For our purposes, we extract the data concerning the municipality in which the employment relationship occurs, the (two-digit) sector of activity (following the NACE Rev. 2 classification), the yearly total employment as measured in terms of full-time equivalent (FTE), and the distribution of employment by five-digit occupational codes, as provided by the International Standard Classification of Occupations 2008 (ISCO08).

The second source of information is the Bureau van Dijk AIDA database which provides the balance sheets of all Italian limited liability companies. Specifically, we extract data related to all manufacturing companies in Emilia-Romagna, active between 2008 and 2017, with positive turnover and non-missing entries for all the variables used in the empirical section such as value-added, material and immaterial assets. This yields a total sample of about 6,000 firms across the whole period of interest.[4]

Data on industrial robots, our third data source, come from the World Robotics 2020 database managed by the International Federation of Robotics (IFR). From this dataset, we get information on the yearly number of installations[5] in Italian manufacturing industries between 2008 and 2017. The sectoral detail of the IFR data is slightly more aggregated than that of the SILER and AIDA; therefore, we pool manufacturing sectors as follows, using the NACE Rev. 2 classification: 10–12 food and beverages; 13–15 textiles; 16 wood and furniture; 17–18 paper; 19 pharmaceutical and cosmetics; 20–21 other chemical products; 22 rubber and plastics products (non-automotive); 23 glass, ceramics, stone, mineral products (non-automotive); 24 basic metals; 25 metal products (non-automotive); 26–27 electrical/electronics; 28 industrial machinery; 29 automotive; 30 other vehicles; 91 all other manufacturing sectors. Since data on the annual flows of robot installations are only available at the country-sector level, to make them vary at the LLMA-sector level in Emilia Romagna we proceed as follows. First, we compute the weight $\omega_{s,\,ER}^{2008}$ of Emilia Romagna (ER)'s sectors s in Italy, using the information on the sector and regional employment from the 'Archivio Statistico delle Imprese Attive' (ASIA): $\omega_{s,ER}^{2008} = \frac{FTE_{s,ER}^{2008}}{FTE_{s,\,Italy}^{2008}}$. To avoid simultaneity with our robotization variables, we compute the weights at the beginning of the available period, 2008. Second, we compute the 2008 weight $\omega_{s,LLMA}^{2008}$ of each manufacturing sector s of every LLMA i in the ER region using information from SILER: $\omega_{s,i,}^{2008} = \frac{FTE_{s,i}^{2008}}{FTE_{s,\,ER}^{2008}}$. Third, we allocate the yearly robot installations ($ROBOT$) available at country-sector level using the two previous weights, as follows: $ROBOT_{s,\,i,\,t} = \omega_{s,ER}^{2008}\,\omega_{s,i}^{2008} ROBOT_{s,\,Italy,\,t}$. Finally, we compute our final robot density variable by dividing $ROBOT_{s,i,t}$ by the number of FTE employees in each sector s and LLMA i in 2008, as provided by SILER, so as to mitigate against potential endogeneity. In this way, we obtain an annual local measure of sector-specific robot density (*ROBDEN*), as follows[6]:

$$ROBDEN_{s,i,t} = \frac{ROBOT_{s,i,t}}{FTE^{2008}_{s,i}}. \tag{1}$$

Table 1 shows the summary statistics of all our main variables and the correlation between the explanatory variables, once accounting for the fixed effects. Figure 1 shows the geography of labour productivity and robot density across the LLMAs of Emilia-Romagna (average 2008–2017). Interestingly, we observe that the distribution of both variables is quite heterogeneous across LLMA: on average, the most central areas of the region, corresponding to the administrative provinces of Bologna, Modena, Parma, and Piacenza are the most productive, while those most exposed to robots are Bologna, Ferrara, and Parma.

Concerning the fourth data source, we retrieve information on occupations, tasks, and skills from the 'Indagine Campionaria sulle Professioni' (ICP), a survey managed by the Italian Institute for Vocational Education and Training (INAPP) and the Italian Statistical Institute (ISTAT). Similarly, to the O*NET survey in the US, the ICP collects information on a sample of roughly 16,000 employees, who answered a series of questions concerning the level and importance of 255 items concerning tasks, skills, work attitudes, and working conditions for each of the approximately 800 five-digit occupational titles identified by the ISCO08 (see Antonietti et al. 2022 for further details). This dataset is here used to compute our routine-task employment variables.

We start by using a series of the 41 working activities (tasks) categorized according to the cross-partition proposed by Autor, Levy, and Murnane (2003) (see Table A.1 in the online Appendix for further details). Using this taxonomy, we calculate the Routine Task Intensity (RTI) index by using the following formula proposed by Autor and Dorn (2013) and Goos, Manning, and Salomons (2014):

$$RTIJ = ln(RMj) + RCj - ln\ [(NRIj) + NRAj + ln(NRMj)] \tag{2}$$

where j is the five-digit occupational title, defined by the ISCO08 classification. The final

Table 1. Summary statistics.

		Mean	Std Dev	Corr with ROBDEN
VA	within		17636.87	0.006**
	between		104397.7	
	overall	39815.06	105758.7	
EMP	within		152.381	0.053**
	between		760.2166	
	overall	336.6763	774.4818	
VA/EMP	within		77.93231	−0.012
	between		80.70797	
	overall	102.8755	112.1261	
ROBDEN	within		342.6856	1
	between		320.5192	
	overall	78.36627	342.6856	
INTK	within		1301.254	−0.0298
	between		1860.042	
	overall	157.7223	2268.276	

Note: Correlations are extracted from a panel fixed-effects regression. ** significant at 5% level; *** significant at 1% level.

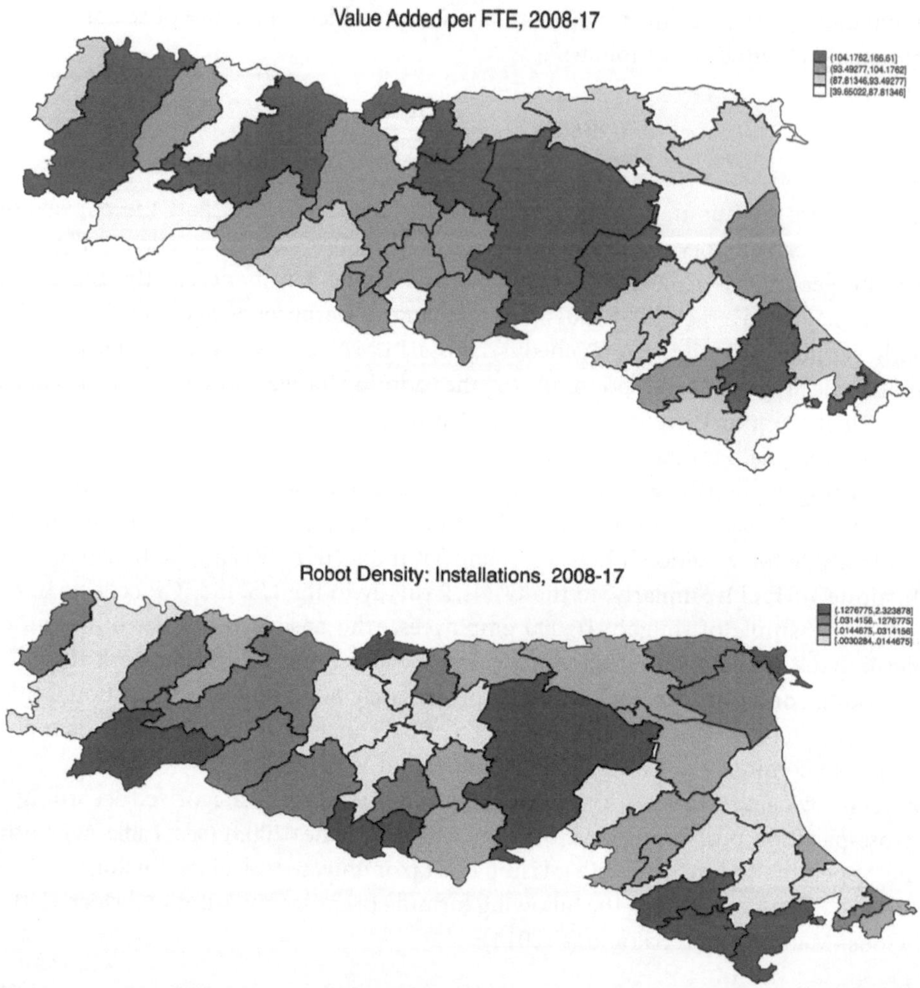

Figure 1. Labour productivity and robot density in Emilia-Romagna. Source: authors' elaborations.

RTI is given by the difference between the natural logarithm of the sum of the average scores of routine tasks (Routine Manual, *RM*, and Routine Cognitive, *RC*) and the sum between the natural logarithm of the sum of the average scores of Non-Routine Interactive (*NRI*), and Non-Routine Analytical, (*NRA*), tasks and the log of the average scores of Non-Routine Manual (*NRM*) tasks for each occupation. As such, it can be interpreted as the ratio between the relevance of the routine task inputs associated with each occupation and its relative non-routine task inputs: the higher the ratio, the more routine tasks are important for that specific occupation, and vice-versa. The RTI in (2) is computed for each of the available occupational titles. Using the SILER and knowing the actual distribution of occupational titles across workers in each LLMA-sector of Emilia-Romagna, we proceed by computing, for every sector and LLMA, the share of employees performing high-routine tasks, namely employees associated to an occupational title j with an average RTI_j higher than the median (*HRTI*). We also consider a stricter version of the routine jobs by computing the share of employees performing very high routine tasks, that is employees whose occupational title j is associated with an average RTI_j belonging to the first quartile of the distribution (*HRTIQ*1).

Our final dataset is made of 3920 observations, obtained from 39 LLMAs x 15 two-digit sectors x 10 years. However, not all the sectors are present, in each year in each LLMA. For this reason, the number of observations is lower than 3920, and reduces to 3096, 2784, and 2437 observations when we include one-year, two-year, and three-year lagged regressors, respectively.

3.2. Econometric strategy

To assess the relationship between the local exposure to robots and labour productivity, we estimate the following first-step model:

$$lnY_{ist} = \beta_1 lnROBDEN_{ist-k} + \beta_2 lnINTK_{ist-k} + \mu_i + \lambda_s + \theta_t + \varepsilon_{ist}, \quad (3)$$

where Y is, respectively, value added per FTE employee (*VA/EMP*), value-added (*VA*) and number of full-time employees (*EMP*) in LLMA i, sector s, and year t; *ROBDEN* is our measure of robot density, computed as number of industrial robot installations per FTE employee; *INTK* is a measure of intangible capital, computed as the yearly value of immaterial[7] assets per 2008 employee; μ_i is a vector of LLMA-specific fixed effects; λ_s is a vector of sector-specific fixed effects; θ_t is a vector of year dummies that are included to capture the evolution of Y in the entire Emilia-Romagna region and across the period of observation; ε_{ijt} is the stochastic error component. We transform all our continuous variables in a natural logarithm to interpret the estimated βs as elasticities. To allow the relationship between robots and productivity/employment to take shape over the years, we set $k = 1$, 2, and 3. We estimate equation (1) using a fixed effects (FE) estimator and we cluster the standard errors at the LLMA-sector level to control for unobserved arbitrary within-group correlation. Our parameter of interest is β_1, which represents the elasticity of labour productivity, and its two constituent elements (value-added and employment), with respect to robot density. Our observations can also be affected by cross-sectional correlation (CD), due to spatial spillovers across LLMAs that can potentially bias the standard errors. For this reason, we also re-estimate equation (3) using the Driscoll and Kraay (1998) methodology, which allows for standard errors that are robust to very general forms of spatial and temporal dependence across units (Hoelche 2007). We always report these CD-robust standard errors in brackets.

As a refinement, when looking at the possible relation between robots and employment, we also consider the routine-task component of the latter. More specifically, we replace Y with the annual share of employees performing routine tasks on the job, as measured by the RTI index described in section 3.1. If robots are more likely to displace workers employed in occupations with relatively higher RTI, we do expect β_1 to be negative and significant.

3.3. First-step results

Table 2 shows the results from the first set of regressions. From Columns 1–3 we find that the estimated coefficient of *ROBDEN* is never statistically significant. From Columns 4–6 we also find that the relationship between robot density and value-added too is not

Table 2. First-stage regressions: robots and labour productivity.

	(1) lnVA/EMP	(2) lnVA/EMP	(3) lnVA/EMP	(4) lnVA	(5) lnVA	(6) lnVA	(7) lnEMP	(8) lnEMP	(9) lnEMP
$\ln ROBDEN_{t-1}$	−0.001			0.020			0.010		
	(0.019)			(0.030)			(0.030)		
	[0.013]			[0.011]			[0.019]		
$\ln ROBDEN_{t-2}$		−0.018			0.033			0.011	
		(0.048)			(0.040)			(0.034)	
		[0.022]			[0.035]			[0.027]	
$\ln ROBDEN_{t-3}$			−0.006			−0.058			−0.074**
			(0.053)			(0.035)			(0.034)
			[0.025]			[0.019]***			[0.017]***
lnINTK	✓	✓	✓	✓	✓	✓	✓	✓	✓
Year FE	✓	✓	✓	✓	✓	✓	✓	✓	✓
Nr. Obs.	3096	2784	2421	3075	2766	2421	3096	2784	2437
Nr. groups	379	379	378	379	379	378	379	379	378
R^2 within	0.009	0.005	0.023	0.008	0.006	0.028	0.019	0.019	0.046
R^2 (OLS)	0.574	0.580	0.613	0.884	0.890	0.899	0.872	0.880	0.886

Note: Cluster-robust (LLMA-industry level) standard errors in parentheses. Driscoll-Kraay standard errors in brackets. $^*p < 0.1$, $^{**}p < 0.05$, $^{***}p < 0.01$.

statistically significant for k equal to 1 and 2, while it becomes negative and significant for $k = 3$ and only when we estimate equation (3) using the Driscoll-Kraay approach. In Columns 7–9 we find a negative and significant relationship between robots and employment at time t-3. In other words, a higher exposure of LLMAs to robotization corresponds to a lower level of FTE employment after three years. This delay should not be surprising, since manufacturing workers, when displaced, are usually initially protected by a series of instruments such as unemployment subsidies and wage compensation schemes. The results from Table 2 pinpoint a situation in which, after 2008, the adoption of industrial robots is not accompanied, on average, by significant labour productivity gains: higher adoption of robot equipment is more likely linked to increases in efficiency, rather than in value creation, probably because of its process innovation nature.

In Table 3, we test whether the results concerning value-added and employment can be affected by endogeneity. In line with Dauth et al. (2021) and Dottori (2021), we use a 2SLS approach in which we instrument ln*ROBDEN* with seven variables that measure robot density between 2008 and 2017 in Finland, France, Germany, Norway, Spain, Sweden, and the UK. The instruments are computed as in equation (1), where we use employment weights in 2008 to match the number of robot installations from the 15 sectors of each of the seven European countries with the sectors of the 39 LLMAs in Emilia-Romagna. Since the number of instruments is larger than the number of endogenous variables, we use the Hansen J statistic to test for overidentification. We also test for the exogeneity of ln*ROBDEN* using a Hausman-style 'difference-in-Hansen' statistic, which comes from estimating two equations, one where *ROBDEN* is treated as endogenous, and one where the same variable is treated as exogenous, and where under the null hypothesis the suspected regressor is considered as exogenous.

From Columns 1 and 2 of Table 3, we find that the estimated coefficient of $\ln ROBDEN_{t\text{-}3}$ is still negative, but not statistically significant. We also find that the set of instruments is strong since the F statistic is much larger than 10. The Hansen J statistic

Table 3. First-stage regressions: robots and labour productivity, IV estimates.

	(1) lnVA	(2) lnEMP
$\ln ROBDEN_{t\text{-}3}$	−0.049	−0.048
	(0.152)	(0.083)
	[0.042]	[0.063]
lnINTK	✓	✓
Year FE	✓	✓
LLMA FE	✓	✓
Industry FE	✓	✓
N. obs	2437	2437
N. groups	379	379
R^2 within	0.028	0.046
R^2 within – 1st stage	0.153	0.153
R^2 (OLS) – centred	0.816	0.886
R^2 1st stage	0.967	0.967
Kleibergen-Paap F	175.90	175.90
Hansen J (p-value)	0.480	0.824
Endogeneity test (p-value)	0.491	0.699

Note: Cluster-robust (LLMA-industry level) standard errors in parentheses. Driscoll-Kraay standard errors in brackets. *$p < 0.1$, **$p < 0.05$, ***$p < 0.01$.

is not statistically significant, meaning that the model is not overidentified. More interestingly, the difference-in-Hansen statistic too is not statistically significant, indicating that our three-year lagged robot density variable can be considered exogenous. In such a case, the (panel) OLS estimates are not only consistent but also more efficient than the instrumental variable ones.

Leveraging on the results of Table 2, Column 9, and Table 3, we further investigate which type of employees are mostly hit by robot exposure. Table 4 shows the results of the estimates of equation (3) when the dependent variable is the share of employees accomplishing high-routine tasks. Column 1 shows a negative and significant (at the 10% level) correlation between ln*ROBDEN* and ln*HRTI*. In Column 3, we find that such a negative correlation holds even when the dependent variable is ln*HRTIQ1*. Columns 2 and 4, instead, show the results from the 2SLS regression. As for Table 3, we still find that the endogeneity test does not reject the null hypothesis, meaning that we can consider $\ln ROBDEN_{t-3}$ as exogenous and rely on the results in Columns 1 and 3.

Taking stock of these results, we proceed with the second step of the analysis, where we investigate the role of similarity and complexity in moderating the relationship between robots and employment.

4. Robots and occupational structure: the mediating role of similarity and complexity

4.1. Theoretical framework

In this section, we propose a novel idea on the possible role that occupation/task similarity and complexity might play in mediating the relationship between robots and employment at the local level and contribute to interpreting the results obtained in the previous Section.

Similarity[8] measures the 'distance' among occupations according to the types of required skills and performed tasks in each occupation (Nedelkoska, Diodato, and Neffke 2018). The lower is such a distance, the higher should be the worker's

Table 4. First-stage regressions: robots and high-routine employment.

	(1) lnHRTI	(2) lnHRTI	(3) lnHRTIQ1	(4) lnHRTIQ1
	FE	IV	FE	IV
$\ln ROBDEN_{t-3}$	−0.010*	−0.017	−0.013*	−0.017
	(0.006)	(0.014)	(0.007)	(0.014)
	[0.005]**	[0.005]***	[0.011]	[0.015]
lnINTK	✓	✓	✓	✓
Year FE	✓	✓	✓	✓
Nr. Obs.	2437	2437	2426	2437
Nr. groups	378	378	378	378
R^2 within	0.008	0.007	0.007	0.006
R^2 (OLS)	0.916	0.916	0.931	0.931
R^2 within – 1st stage		0.153		0.153
R^2 1st stage		0.967		0.967
Kleibergen-Paap F		175.90		173.14
Hansen J (*p*-value)		0.839		0.302
Endogeneity test (*p*-value)		0.805		0.870

Note: Cluster-robust (LLMA-industry level) standard errors in parentheses. Driscoll-Kraay standard errors in brackets. $^{*}p < 0.1$, $^{**}p < 0.05$, $^{***}p < 0.01$.

transferability across occupations. The underlying idea is that job reallocation is easier if skills and tasks are similar (but not identical) across workplaces because of the lower costs of training or retraining (Gathmann and Schönberg 2010; Nedelkoska, Diodato, and Neffke 2018; Vona 2021). Occupational similarity may also entail a higher similitude of the knowledge bases of organizations, i.e. a lower cognitive distance (Balland 2012) among them. This would indicate a higher probability of knowledge spillovers, even through workers mobility. Possible measures of similarity include the Euclidean distance (Poletaev and Robinson 2008), the angular separation measure (Gathmann and Schönberg 2010), and the Jaccard similarity index (Nedelkoska, Diodato, and Neffke 2018). According to this framework, similarity provides conceptual and methodological support to the theoretical assumption that workers can easily move from one industry to another if these two industries draw upon similar skills or tasks (Neffke and Henning 2013). Within each region, a certain industry A could thus benefit from a potential inflow of workers exiting an industry B characterized by an occupational structure that fits the skills and tasks of industry A.

A consistent hypothesis with this framework is that the efficiency of human-robot interaction could be boosted if workers' reallocation occurs between industries with similar occupational spaces. This productivity-enhancing effect would stem from two mechanisms operating within each 'industry-local labour market' combination (Farinha et al. 2019): skill complementarity and task interdependence. The impact of robots on productivity would be successfully mediated by inter-industry similarity: robot adoption would not only stimulate a routine-biased task replacement effect, but it would also facilitate workers' reallocation towards occupationally similar sectors characterized by other technological paradigms.

A second aspect that can mediate the robot-employment nexus at the local level is occupational complexity. Differently from relatedness, which can predict the 'easiness' to move from a set of activities/occupations to another one, complexity is considered a proxy of the value of the same set of activities/occupations and has been used to analyse the geography of economic activities from many perspectives (Hidalgo 2021). Notably, when applied to occupational space, complexity can have a twofold role. First, the concept of complexity-biased technical change (CBTC) assumes that workers are allocated across occupations on the basis of their skill endowment and corresponding capabilities to accomplish complex tasks (Caines, Hoffmann, and Kambourov 2017). In a CBTC equilibrium, therefore, the demand for complex tasks will increase with the diffusion of new technologies at the expense of less complex tasks with a substantial benefit in terms of competitiveness of countries and regions in the market of tradable goods, typically associated with the manufacturing industry (Lo Turco and Maggioni 2020). The net employment impact of robot adoption would thus depend on the composition of the three effects identified by Acemoglu and Restrepo (2019): productivity effect, displacement effect, and reinstatement effect. All of them, in fact, are partially incorporated into the notion of job complexity (Antonietti et al. 2022), leaving open the question of whether employment substitution could be offset by the broader range of required tasks engendered by the reinstatement effect. Another issue rests on the way we conceive occupational complexity, i.e. as a combination of a wide variety of occupational profiles (or tasks) and the presence of occupations (or tasks) that are rare, or exclusive, to the local labour market. If a higher adoption of automation technology exposes high-routine employees to displacement, the possibility to find a new job in the manufacturing industry will be harder in highly complex local markets because of a possible mismatch between the skills acquired by the displaced employee and the ones required by new occupations (and possibly induced by the adoption of advanced manufacturing technology, as in Antonietti et al. 2022), high search costs in areas where jobs are highly differentiated and exclusive, and lower and lower demand for routine skills in highly complex and dynamic systems (Modestino, Shoag, and Ballance 2016, 2020). In this perspective, as documented by Neffke, Nedelkoska, and Wiederhold (2022) for Germany, the long-run consequences of this mismatch can take the form of relevant earning losses for the individuals. This idea is also consistent with those studies having shown that occupational complexity is positively associated with product complexity (Lo Turco and Maggioni 2020) in manufacturing industries. A highly complex local production structure (including for instance, aeronautics or pharmaceutical clusters), in fact, may represent a constraint for displaced workers that do not possess those skills and knowledge required by complex products, such as STEM and critical thinking.

For these reasons, we expect that the negative relation between robots and employment holds specifically in LLMA characterized by low similarity and high complexity, whereas we do expect that the negative impact of robots on employment vanishes or becomes less harsh when industries in an LLMA show a high degree of similarity and a low level of job complexity.

As a by-product of the previous reasoning, it is worthwhile investigating whether the negative link between robot adoption and employment is lighter or heavier within industrial districts. This latter corresponds to an LLMA with a strong manufacturing specialization and a high concentration of micro and small enterprises engaged in the same value chain. In this respect, a first simple assumption is that highly specialized LLMAs with many routine manual jobs suffer the highest exposure to robot substitution. However, because of their peculiar characteristics, industrial districts should be also characterized by a high degree of occupation/task similarity across industries and low job complexity. For these reasons, we do expect the negative impact of robots on employment found in Table 2 to vanish within industrial districts.

4.2. Measuring similarity and complexity

In this section, we describe the methodology adopted to compute our measures of similarity and complexity. We first assess the inter-industry occupational similarity (sim) of each LLMA i in 2008, by considering the manufacturing sectors employing at least one FTE and the distribution of occupational titles therein. Occupational titles, o, are identified by three-digit ISCO08 codes, each of which encompasses a specific vector of skills and tasks. To compute our similarity index, we use the 'minimum distance', in terms of o, between sector n and sector s, by subtracting from 1 the weighted Euclidean distance d from n to s.[9] For each couple of sectors, the corresponding formula is the following:

$$sim_{n,s} = 1 - d_{n,s} = 1 - \frac{\sum_{o}^{O} \sqrt{(FTE_{s,o} - FTE_{n,0})^2}}{O} \tag{4}$$

for each LLMA i,

where *%FTE* is the share of FTE employed with occupational title *o* in sector *n*, or *s*, in the LLMA *i*, and *O* is the total number of occupational titles in the same LLMA. This measure is calculated for each couple of sectors and then averaged for each sector n employing at least 1 FTE in *i* through an arithmetic mean, as follows:

$$SIMIL_s = \frac{\sum_n^N sin_{n,s}}{N-1} \tag{5}$$

where $SIMIL_s$ is the inter-industry occupational similarity within each LLMA *i*, while *N* (or *S*) is the number of sectors employing at least one FTE in LLMA *i*. The higher the index, the higher is the average occupational similarity of each sector in each LLMA with respect to all the others, which means that employees share a similar set of skills and tasks.

On top of this, we calculate the scores of occupational ($COMPLEX_{OCC}$) and task complexity ($COMPLEX_{TASK}$), for each LLMA *i*, following the method of reflections developed by Hidalgo and Hausmann (2009) to calculate the Economic Complexity Index (ECI). Accordingly, we first define the Revealed Comparative Advantage (RCA) matrix for each LLMA with respect to occupations O, or tasks T, in 2008. To do so, we use the Balassa index which takes the value of 1 when an LLMA has a comparative advantage in a type of occupation *o*, or task *t*. The RCA is then used to define a binary matrix O(T) where cells are equal to 1 if the corresponding RCA is higher or equal than 1 and 0 otherwise. Once defined this matrix, we measure the diversity and ubiquity of occupations (tasks) by summing over the rows or columns of each matrix, and jointly define the diversity and the ubiquity of each occupation (task), as follows, respectively:

$$Diversity_{o(T)} = k_{i0(T)} = \sum_{o}^{O} O(T)_{io} \tag{6}$$

$$Ubiquity_{o(T)} = k_{o(T)0} = \sum_{i}^{I} O(T)_{io}. \tag{7}$$

The complexity indicator is then obtained through an iterative method of reflections, which corresponds to finding the eigenvalues of a normalized similarity matrix as the following one:

$$\tilde{O}\,(T)_{ii'} = \frac{1}{k_{i0}} \sum_{o}^{O} \frac{O\,(T)_{io(t)} O\,(T)_{i'm}}{k_{o0}} \tag{8}$$

where the rows sum to 1 (see Antonietti et al. 2022 for further details). The *COMPLEX-OCC* (*COMPLEXTASK*) index corresponds to the second largest right eigenvalue of the above similarity matrix, after being standardized as follows:

$$COMPLEX_{OCC(TASK)}i = \frac{\vec{K} - \langle \vec{K} \rangle}{std\,(\vec{K})} \tag{9}$$

where $\underline{\vec{K}}$ is the eigenvector of the matrix associated with the second largest eigenvalues. The higher the index, the higher the average occupational (task) complexity of the LLMA in 2008.

5. Empirical analysis: step 2

5.1. Descriptive statistics

Table 5 shows the average levels of value-added per employee, employment, and robot density by degree (high/low) of similarity (*SIMIL*), occupational complexity ($COMPLEX_{OCC}$), task complexity ($COMPLEX_{TASK}$) and industrial district (ID) status. Interestingly, we observe that LLMAs where the average occupational similarity is higher than the median in 2008 (*High SIMIL*), have significantly higher labour productivity, employment, and robot density than LLMAs with lower than the median average similarity (*Low SIMIL*). For what concerns complexity, instead, the picture is more nuanced: while labour productivity and employment are higher in high-complexity LLMAs, the exposure to robots is lower. Finally, we find that productivity, employment, and robot density are all lower in industrial districts as compared to the rest of the LLMAs.

Table 6, instead, shows the average value of similarity and complexity for LLMAs identified as industrial districts as compared to non-industrial districts. In addition, we also consider the five main specializations of IDs in Emilia-Romagna, such as textile, manufacturing of household goods, food products, mechanics, and manufacturing of metal products, and we define a dummy variable for each of them. Quite interestingly, we find that the average level of occupational similarity is slightly lower for ID than for non-ID areas, although ID specialized in the manufacturing of mechanical products show the highest degree of similarity across industries. The evidence is much more striking when looking at the average level of occupational and task complexity of ID, which is much lower than that of non-ID areas. This is in line with our expectations: the high level of specialization and homogeneity of occupations within ID contrasts with the high diversity and exclusivity of jobs in non-district, or more urbanized, areas of the region. For this reason, again, we do expect the negative impact of robots on manufacturing employment to be lower in IDs.

5.2. Econometric strategy

As a second step, we test whether the relationship between robot density and employment is mediated by, respectively, the level of industry similarity and occupational (or task) complexity within each LLMA. To do so, we re-estimate equation (3) on different sub-samples, identified according to the degree of industry similarity and

Table 5. Productivity, employment, and robot density by similarity, complexity, and district status.

	VA/EMP	EMP	ROBDEN
High SIMIL	113.05	471.74	56.81
Low SIMIL	92.69***	201.61***	99.91***
High $COMPLEX_{OCC}$	112.35	453.11	45.70
Low $COMPLEX_{OCC}$	93.77***	224.90***	109.73***
High $COMPLEX_{TASK}$	109.82	447.15	50.73
Low $COMPLEX_{TASK}$	95.87***	225.07***	106.29***
ID	97.94	304.39	49.74
NO ID	105.71*	355.22*	94.81***

Note: Cells report the t-test on the differences in mean. *Significant at 10% level; ** t-test significant at 5% level; *** t-test significant at 1% level.

Table 6. Similarity and complexity within and outside industrial districts.

	SIMIL	$COMPLEX_{OCC}$	$COMPLEX_{TASK}$
NO ID	0.541	0.391	0.125
ID	0.512	−0.058	−0.387
- ID TEXTILE	0.361	2.044	0.852
- ID HOUSE	0.438	−0.549	−0.805
- ID FOOD	0.505	−0.341	−0.659
- ID MECH	0.554	−0.115	−0.293
- ID METAL	0.508	−0.584	−1.292

occupational (task) complexity of LLMAs. Specifically, we define high-similarity if, in 2008, the degree of SIMIL of an industry in a LLMA is higher than the median, and low-similarity if, in 2008, the degree of *SIMIL* is lower than the median. Similarly, we define high-complexity if, in 2008, the average degree of $COMPLEX_{OCC}$ (or $COMPLEX_{TASK}$) is higher than the median, and low-complexity otherwise. As a sensitivity test, we define high/low-*SIMIL Q*1 and high/low-$COMPLEX_{OCC}$ *Q*1 using the threshold value of the first quartile of the corresponding distributions instead of the median.

In an additional specification, we combine high/low similarity and high/low complexity, and we re-estimate equation (3) on the samples of high-*SIMIL*/low-$COMPLEX_{OCC}$ and low-*SIMIL*/high-$COMPLEX_{OCC}$ LLMAs. As before, we test for the sensitivity of the results using the first quartile instead of the median to define high/low-similarity and complexity local labour markets.

We also regress equation (3) on the sample of LLMAs defined as industrial districts (ID) as compared to non-industrial district LLMAs (NO ID). According to our expectations, the negative correlation between *ROBDEN* and *EMP* should vanish within IDs, while remaining significant in non-district areas. Finally, we also split the ID sample into five sub-samples, each identified on the main specialization of the ID (i.e. textile, household goods, food, mechanics, and metal products) and we re-estimate equation (3) on each of them, as compared to the NO ID sample.

5.3. Results: step 2

The results of the step 2 regressions are provided in Tables 7–9. From Table 7, Columns 1 and 2, we find that the negative correlation between ln$ROBDEN_{t-3}$ and ln*EMP* is statistically significant only in an LLMA characterized by a low level of inter-industry occupational similarity, whereas the significance disappears in high-similarity LLMAs. As expected, LLMAs, where similarity is high, do not seem to suffer any significant manufacturing employment loss after their exposure to robots, while in LLMAs where similarity is below the median a 10% increase in robot density corresponds to an average 1.1% decrease in FTE employment. Results are confirmed in Columns 3 and 4 where we define high and low similarity areas on the base of the first quartile of the SIMIL distribution. Still, in line with our hypotheses, we also find that a higher than the median degree of occupational complexity positively mediates the negative correlation between robots and employment: from Column 5, it can be observed that a 10% increase in *ROBDEN* corresponds to an average − 0.92% (significant at the 10% level) in *EMP* when $COMPLEX_{OCC}$ is high, whereas in Column 6 the estimated coefficient of

Table 7. Second-stage regressions: robots and employment.

	(1) High SIMIL	(2) Low SIMIL	(3) High SIMIL Q1	(4) Low SIMIL Q1	(5) High $COMPLEX_{OCC}$	(6) Low $COMPLEX_{OCC}$	(7) High $COMPLEX_{OCC}$ Q1	(8) Low $COMPLEX_{OCC}$ Q1
$lnROBDEN_{t-3}$	−0.044	−0.110**	−0.034	−0.089**	−0.092*	−0.052	−0.102*	−0.064
	(0.048)	(0.049)	(0.068)	(0.040)	(0.051)	(0.044)	(0.064)	(0.040)
	[0.030]	[0.026]***	[0.059]	[0.018]***	[0.026]***	[0.021]**	[0.030]***	[0.022]***
lnINTK	✓	✓	✓	✓	✓	✓	✓	✓
Year FE	✓	✓	✓	✓	✓	✓	✓	✓
Nr. Obs.	1240	1197	628	1809	1219	1197	660	1777
Nr. groups	193	185	96	282	189	185	103	275
R^2 within	0.069	0.031	0.082	0.037	0.042	0.060	0.049	0.054
R^2 (OLS)	0.872	0.885	0.865	0.892	0.881	0.889	0.870	0.886
Endog. (p)	0.966	0.835			0.919	0.723		

Note: Cluster-robust (LLMA-industry) standard errors in parentheses. Driscoll-Kraay standard errors in brackets. $^{*}p < 0.1$, $^{**}p < 0.05$, $^{***}p < 0.01$.

ROBDEN is not statistically different from zero. When we use the Driscoll and Kraay approach, we find that the negative coefficient of ln*ROBDEN* in low-complexity LLMAs turns significant at 5% level, but the value of the elasticity is almost half of that estimated in Column 5 for high-complexity LLMAs. Similar results emerge in Columns 7 and 8, and when we replace $COMPLEX_{OCC}$ with $COMPLEX_{TASK}$ (see online Appendix, Table A.2).

This evidence is further strengthened in Table 8, where we estimate the robot-employment relationship in two opposite scenarios: low similarity/high complexity versus low similarity/high complexity. The results in Columns 1 and 2 confirm that the only situation in which we find a negative and (strongly) significant coefficient is that of LLMAs characterized by a low occupational similarity and high occupational complexity.

Finally, Table 9 provides the results concerning the role of IDs (See Table A.3 in the online Appendix for the results of the regression analysis based on interaction variables). As expected, Column 1 shows that the estimated coefficient of robot density remains negative and statistically significant only outside IDs. We find that a 10% increase in robot exposure corresponds to an average 0.92% decrease in employment. Again, when we correct the standard errors for CD, the coefficient of ln*ROBDEN* within IDs turns statistically significant at 5% level, but its magnitude is much lower than that for non-ID areas. Looking back at Table 6, the main reason can be attributed to the low level of occupational complexity that governs IDs as compared to the other areas. Interestingly, Columns 3–7 show that, when we distinguish the IDs by their specialization, a higher exposure to robotization corresponds to a significant decrease in employment only when we adjust the standard errors to account for CD and only in IDs specialized textile, food, and mechanics. In IDs specialized in home goods manufacturing the impact of robotization on employment is even positive and significant, whereas in metal manufacturing is positive but not statistically significant. In Column 3, we find that a 10% increase in robot density corresponds to an average 5.5% decrease in employment. Such a high elasticity, however, is not surprising: if we look again at Table 6, we note that textile IDs are those characterized by the lowest degree of similarity and the highest levels of occupational and task complexity.

Table 8. Second-stage regressions: robots and employment, combining similarity and complexity.

	(1) Low SIMIL High $COMPL_{OCC}$	(2) High SIMIL Low $COMPL_{OCC}$	(3) Low SIMIL Q1 High $COMPL_{OCC}$ Q1	(4) High SIMIL Q1 Low $COMPL_{OCC}$ Q1
$\ln ROBDEN_{t-3}$	−0.220**	−0.092	−0.152**	−0.071
	(0.075)	(0.070)	(0.072)	(0.076)
	[0.042]***	[0.050]*	[0.049]***	[0.051]
lnINTK	✓	✓	✓	✓
Year FE	✓	✓	✓	✓
Nr. Obs.	591	612	559	527
Nr. groups	91	95	87	80
R^2 within	0.058	0.110	0.059	0.094
R^2 (OLS)	0.858	0.840	0.870	0.837
Endogeneity	0.107	0.476	0.588	0.862

Note: Cluster-robust (LLMA-industry) standard errors in parentheses. Driscoll-Kraay standard errors in brackets. *$p < 0.1$, **$p < 0.05$, ***$p < 0.01$.

Table 9. Second-stage regressions: the moderating role of industrial districts.

	(1) No district	(2) District	(3) Textile	(4) Home	(5) Food	(6) Mech	(7) Metals
$\ln ROBDEN_{t-3}$	−0.092**	−0.042	−0.552	0.144	−0.146	−0.075	0.163
	(0.045)	(0.054)	(0.359)	(0.114)	(0.205)	(0.079)	(0.294)
	[0.028]***	[0.016]**	[0.120]***	[0.062]**	[0.023]***	[0.026]***	[0.122]
lnINTK	✓	✓	✓	✓	✓	✓	✓
Year FE	✓	✓	✓	✓	✓	✓	✓
Nr. Obs.	1564	873	62	114	1219	417	44
Nr. groups	241	137	12	20	18	79	8
R^2 within	0.046	0.056	0.208	0.034	0.124	0.082	0.300
R^2 (OLS)	0.866	0.885	0.737	0.919	0.891	0.887	0.806
Endog. (p)	0.788	0.588		0.171			

Note: Cluster-robust (LLMA-industry) standard errors in parentheses. Driscoll-Kraay standard errors in brackets. *$p < 0.1$, **$p < 0.05$, ***$p < 0.01$.

6. Conclusions

This paper investigates a still overlooked aspect of the impact that industrial robots can have on labour productivity and employment at the LLMA-industry level: the role of occupational similarity and complexity. To do so, we use a longitudinal dataset merging information on workers allocation by industry, occupation, and LLMA in the Emilia-Romagna region during the period 2008–2017, with O*NET-type survey data on the distribution of tasks by occupational title, and with international data on the actual installations of industrial robots. Our empirical analysis develops along two steps. In the first, our estimates show that higher robot exposure does neither increase labour productivity nor manufacturing value-added, but, rather, it corresponds to declining (high routine) full-time employment after three years. This delayed effect is probably due to the presence of temporary labour protection and wage compensation schemes (for example the 'Cassa Integrazione' in Italy). Instead, the task reallocation from workers to robots can be substantial, implying a decrease in the labour share in industry value-added. The second step deals with the characteristics of the occupational space of each LLMA. Co-location of industries with similar occupations, or that share similar tasks, combined with the presence of low complex jobs, mitigates labour displacement. The underlying logic is that workers displaced by robots can more easily move to other manufacturing sectors if the latter share similar tasks and the overall occupational space of the area is not too complex, i.e. not too diversified and characterized by exclusive (and less ubiquitous or repeatable) jobs.

This result indicates that local systems of production can take two distinct trajectories to deal with the exposure of their manufacturing sectors to automation technology. The first is one that preserves jobs, and the underlying competencies, through the transfer of workers to less robotized manufacturing industries. Such a trajectory would replicate the smart specialization model in the occupational space: less robotized sectors may benefit from the skills and capabilities acquired by workers previously employed in more robotized (but occupationally 'similar') industries. This might come at the cost of lower aggregate productivity caused by the absence of local synergies deriving from the co-location of different and complementary jobs as posited by a CBTC approach (Farinha et al. 2019). The second trajectory, on the contrary, is one characterized by employment substitution, slight productivity growth, and highly complex occupational spaces. It could be that productivity increases will lead to future growth of labour demand due to an expansion of total output (Acemoglu and Restrepo 2019). However, workers displaced by robotization could face large and persistent earnings losses, unless they participate in an upskilling/reskilling process (Neffke, Nedelkoska, and Wiederhold 2022). Moreover, local systems experiencing such displacement may face a stronger decline of its stock of specific human capital (Poletaev and Robinson 2008). In this regard, one can hypothesize the existence of a trade-off between two sources of productivity gains: robotization-driven labour substitution and intra-LLMA knowledge spillovers engendered by workers' mobility.

In this framework, we have found a possible role for industrial districts too. Being characterized by an average level of cross-industry occupational similarity, but a low degree of complexity (except for those specialized in textile manufacturing), industrial districts might play the role of 'shock absorbers' in the region. When displaced by robots, (routine) manufacturing employees can find alternative job opportunities more easily because of the presence of similar industries and less complex occupational profiles. This poses an interesting challenge with respect to the role of specialization and diversification for the development of manufacturing regions. The main policy implications deal with the potential role of regional agencies, trade unions and industrial associations in favouring inter-industry workers' reallocation within LLMAs. Not only could they support displaced workers in finding jobs that are close to their previous occupation, but they may also provide workers with industry-specific skills referred to those sectors that are more 'similar', in terms of occupations, to the ones mostly exposed to robotization. This 'mitigating' policy could be accompanied by training policies aimed at broadening the set of tasks that workers can accomplish in order to accompany the demand for increasingly complex tasks entailed by robotization.

A limit of this paper is that it focuses on one region only, Emilia-Romagna. Extending the analysis to the entire Italian context, or even to European regions, will represent an intriguing avenue for future research. Further research could also be oriented to broaden the analysis to the impact of industrial robots on productivity and employment of the local service industries, which can benefit from positive spillovers from manufacturing. In addition, the analysis could be extended to other technologies, such as Artificial Intelligence or Internet of Things, as it is reasonable to argue that the impact of Industry 4.0 technologies on employment and productivity could be different from that of industrial robots.

Notes

1. Available here: https://morningconsult.com/2021/06/21/automation-workforce-jobs-polling/.
2. See: https://ec.europa.eu/eurostat/statistics-explained/index.php?title=Labour_market_statistics_at_regional_level#Employment.
3. Occupations are commonly used as a proxy for the skills and abilities of the employees beyond their formal education (Autor, Levy, and Murnane 2003; Bacolod, Blum, and Strange 2009; Andersson, Klaesson, and Larsson 2014). Accordingly, occupations are defined by a set of tasks, skills, education, and attitudes needed to perform the work. The occupational space of an entity, such as firm, industry or region, reflects the degree of

variety/similarity of skills or tasks that characterize such an entity (Muneepeerakul et al. 2013).

4. For an in-depth discussion of the use of the AIDA Bureau van Dijk archive for economic analyses in the Italian context, see Grazzi, Piccardo, and Vergari (2019).
5. We choose to use the yearly installations rather than the operational stock because our dependent variable, value-added per employee, is a flow measure.
6. Our variable slightly differs from Dauth et al. (2021) and Dottori (2021), who compute robot density at the LLMA level. Differently, we want to keep a sectoral variability to control for sector-specific fixed effects in the regressions and to match it with our similarity measure, which varies at sector and LLMA level.
7. We do not include a variable for tangible capital, as given by the yearly value of material assets, because it can partially overlap with our robot density variable. In any case, the results do not change if we add the value of tangible capital among the regressors.
8. Relatedness metrics better captures intra-industry occupational closeness because it is a hierarchical measure that can be used to deepen the occupational space within a single industry, while similarity is more suitable for inter-industry occupational matching because it refers to the occupational distance between each couple of sectors. We choose similarity for our analysis because robot density per sector is measured at 2-digit level and thus we cannot look at intra-industry heterogeneity in terms of robotization.
9. Since the data from SILER are anonymised with respect to the individual worker identifier, we cannot apply the methodology developed by Neffke and Henning (2013) to compute skill relatedness.

Disclosure statement

No potential conflict of interest was reported by the author(s).

ORCID

Roberto Antonietti http://orcid.org/0000-0002-2172-4062

References

Acemoglu, D., C. Lelarge, and P. Restrepo. 2020. “Competing with Robots: Firm-Level Evidence from France.” *AEA Papers and Proceedings* 110: 383–388. doi:10.1257/pandp.20201003

Acemoglu, D., and P. Restrepo. 2018. “The Race Between Man and Machine: Implications of Technology for Growth, Factor Shares, and Employment.” *American Economic Review* 108 (6): 1488–1542. doi:10.1257/aer.20160696

Acemoglu, D., and P. Restrepo. 2019. “Automation and New Tasks: How Technology Displaces and Reinstates Labor.” *Journal of Economic Perspectives* 33 (2): 3–30. doi:10.1257/jep.33.2.3

Acemoglu, D., and P. Restrepo. 2020. “Robots and Jobs: Evidence from US Labor Markets.” *Journal of Political Economy* 128 (6): 2188–2244. doi:10.1086/705716

Acemoglu, D., and P. Restrepo. 2021. “Tasks, Automation, and the Rise in US Wage Inequality.” National Bureau of Economic Research, Working Paper N. 28920.

Andersson, M., J. Klaesson, and J. Larsson. 2014. “The Sources of the Urban Wage Premium by Worker Skills: Spatial Sorting or Agglomeration Economies?” *Papers in Regional Science* 93 (4): 727–747. doi:10.1111/pirs.12025

Antonietti, R., L. Cattani, F. Gambarotto, and G. Pedrini. 2022. “Education, Routine, and Complexity-Biased Key Enabling Technologies: Evidence from Emilia-Romagna, Italy.” *Industry and Innovation*, doi:10.1080/13662716.2022.2045910.

Autor, D., and D. Dorn. 2013. “The Growth of low-Skill Service Jobs and the Polarization of the US Labor Market.” *American Economic Review* 103 (5): 1553–1597. doi:10.1257/aer.103.5.1553.

Autor, D., F. Levy, and R. Murnane. 2003. "The Skill Content of Recent Technological Change: An Empirical Exploration." *The Quarterly Journal of Economics* 118 (4): 1279–1333. doi:10.1162/003355303322552801.

Bacolod, M., B. Blum, and W. Strange. 2009. "Urban Interactions: Soft Skills Versus Specialization." *Journal of Economic Geography* 9 (2): 227–262. doi:10.1093/jeg/lbn057

Balland, P. 2012. "Proximity and the Evolution of Collaboration Networks: Evidence from Research and Development Projects Within the Global Navigation Satellite System (GNSS) Industry." *Regional Studies* 46 (6): 741–756. doi:10.1080/00343404.2010.529121

Balland, P., and D. Rigby. 2017. "The Geography of Complex Knowledge." *Economic Geography* 93 (1): 1–23. doi:10.1080/00130095.2016.1205947.

Bessen, J. 2019. Automation and Jobs: When Technology Boosts Employment, Boston Univ. School of Law, Law and Economics Research Paper N. 17-09.

Boix, R., and J. Trullen. 2010. "Industrial Districts, Innovation and I-District Effect: Territory or Industrial Specialization?" *European Planning Studies* 18 (10): 1707–1729. doi:10.1080/09654313.2010.504351.

Boschma, R., R. Eriksson, and U. Lindgren. 2008. "How Does Labour Mobility Affect the Performance of Plants? The Importance of Relatedness and Geographical Proximity." *Journal of Economic Geography* 9 (2): 169–190. doi:10.1093/jeg/lbn041.

Brusco, S. 1982. "The Emilian Model: Productive Decentralisation and Social Integration." *Cambridge Journal of Economics* 6 (2): 167–184.

Brynjolfsson, E., and A. McAfee. 2012. *Race Against the Machine: How the Digital Revolution Is Accelerating Innovation, Driving Productivity, and Irreversibly Transforming Employment and The Economy*. Lexington, MA: Digital Frontier Press.

Caines, C., F. Hoffmann, and G. Kambourov. 2017. "Complex-Task Biased Technological Change and the Labor Market." *Review of Economic Dynamics* 25: 298–319. doi:10.1016/j.red.2017.01.008.

Carbonero, F., E. Ernst, and E. Weber. 2020. "Robots Worldwide: The impact of Automation on Employment and Trade." IAB Discussion Papers N. 7|2020.

Dauth, W., S. Findeisen, J. Suedekum, and N. Woessner. 2021. "The Adjustment of Labor Markets to Robots." *Journal of the European Economic Association* 19 (6): 3104–3153. doi:10.1093/jeea/jvab012

De Vries G., Gentile E., Miroudot S., and K. Wacker. 2020. "The Rise of Robots and the Fall of Routine Jobs." *Labour Economics* 66: 101885. doi:10.1016/j.labeco.2020.101885

Dixon, J., B. Hong, and L. Wu. 2021. "The Robot Revolution: Managerial and Employment Consequences for Firms." *Management Science* 67 (9): 5586–5605. doi:10.1287/mnsc.2020.3812

Dottori, D. 2021. "Robots and Employment: Evidence from Italy." *Economia Politica* 38 (2): 739–795. doi:10.1007/s40888-021-00223-x

Driscoll, J., and A. Kraay. 1998. "Consistent Covariance Matrix Estimation with Spatially Dependent Panel Data." *Review of Economics and Statistics* 80: 549–560. doi:10.1162/003465398557825

Farinha, T., P. Balland, A. Morrison, and R. Boschma. 2019. "What Drives the Geography of Jobs in the US? Unpacking Relatedness." *Industry and Innovation* 26 (9): 988–1022. doi:10.1080/13662716.2019.1591940

Frenken, K., F. van Oort, and T. Verburg. 2007. "Related Variety, Unrelated Variety and Regional Economic Growth." *Regional Studies* 41 (5): 685–697. doi:10.1080/00343400601120296.

Frey C., and Osborne M. 2017. "The Future of Employment: How Susceptible are Jobs to Computerisation?" *Technological Forecasting and Social Change* 114: 254–280. doi:10.1016/j.techfore.2016.08.019

Gathmann, C., and U. Schönberg. 2010. "How General is Human Capital? A Task-Based Approach." *Journal of Labor Economics* 28 (1): 1–49. doi:10.1086/649786

Gentili, A., F. Compagnucci, M. Gallegati, and E. Valentini. 2020. "Are Machines Stealing Our Jobs?" *Cambridge Journal of Regions, Economy and Society* 13 (1): 153–173. doi:10.1093/cjres/rsz025

Goos, M., A. Manning, and A. Salomons. 2014. "Explaining Job Polarization: Routine-Biased Technological Change and Offshoring." *American Economic Review* 104 (8): 2509–2526. doi:10.1257/aer.104.8.2509.

Graetz, G., and G. Michaels. 2018. "Robots at Work." *The Review of Economics and Statistics* 100 (5): 753–768. doi:10.1162/rest_a_00754.

Grazzi, M., C. Piccardo, and C. Vergari. 2019. "Building a Firm Level Dataset for the Analysis of Industrial Dynamics and Demography." *Journal of Economic and Social Measurement* 43 (3-4): 169–197. doi:10.3233/JEM-180456

Hidalgo, C. 2021. "Economic Complexity Theory and Applications." *Nature Reviews Physics* 3 (2): 92–113. doi:10.1038/s42254-020-00275-1

Hidalgo, C., and R. Hausmann. 2009. "The Building Blocks of Economic Complexity." *Proceedings of the National Academy of Sciences* 106 (26): 10570–10575. doi:10.1073/pnas.0900943106

Hidalgo, C., et al. 2018. "The Principle of Relatedness." In *Unifying Themes in Complex Systems IX. ICCS 2018. Springer Proceedings in Complexity*, edited by A. Morales, C. Gershenson, D. Braha, A. Minai, and Y. Bar-Yam, 451–457. Cham: Springer.

Hoelche, D. 2007. "Robust Standard Errors for Panel Regressions with Cross-Sectional Dependence." *The Stata Journal: Promoting Communications on Statistics and Stata* 7 (3): 281–312. doi:10.1177/1536867X0700700301

Klenert, D., E. Fernandez-Macias, and J. Antón. 2020. "Do Robots Really Destroy Jobs? Evidence from Europe." *Economic and Industrial Democracy*, doi:10.1177/0143831X211068891.

Koch, M., I. Manuylov, and M. Smolka. 2021. "Robots and Firms." *The Economic Journal* 131 (638): 2553–2584. doi:10.1093/ej/ueab009

Kromann, L., N. Malchow-Møller, J. Skaksen, and A. Sørensen. 2020. "Automation and Productivity—A Cross-Country, Cross-Industry Comparison." *Industrial and Corporate Change* 29 (2): 265–287. doi:10.1093/icc/dtz039

Lo Turco, A., and D. Maggioni. 2020. "The Knowledge and Skill Content of Production Complexity." *Research Policy*, doi:10.1016/j.respol.2020.104059.

McKinsey Global Institute. 2017. *A Future That Works. Automation, Employment, and Productivity*. New York: McKinsey Global Institute.

Modestino, A., D. Shoag, and J. Ballance. 2016. "Downskilling: Changes in Employer Skill Requirements Over the Business Cycle." *Labour Economics* 41: 333–347. doi:10.1016/j.labeco.2016.05.010

Modestino, A., D. Shoag, and J. Ballance. 2020. "Upskilling: Do Employers Demand Greater Skill When Workers are Plentiful?" *The Review of Economics and Statistics* 102 (4): 793–805. doi:10.1162/rest_a_00835

Muneepeerakul, R., J. Lobo, S. Shutters, A. Goméz-Liévano, and M. Qubbaj. 2013. "Urban Economies and Occupation Space: Can They Get "There" from "Here"?" *PloS ONE* 8 (9): e73676. doi:10.1371/journal.pone.0073676

Nedelkoska, L., D. Diodato, and F. Neffke. 2018. "Is Our Human Capital General Enough to Withstand the Current Wave of Technological Change?" CID Research Fellow and Graduate Student Working Paper Series N. 2018.93. Cambridge: Harvard University.

Nedelkoska, L., and F. Neffke. 2019. "Skill Mismatch and Skill Transferability: Review of Concepts and Measurements." Papers in Evolutionary Economic Geography 19-21, Utrecht University.

Neffke, F., and M. Henning. 2013. "Skill Relatedness and Firm Diversification." *Strategic Management Journal* 34 (3): 297–316. doi:10.1002/smj.2014

Neffke, F., L. Nedelkoska, and S. Wiederhold. 2022. "Skill Mismatch and the Costs of Job Displacement." CESifo Working Papers N. 9703.

OECD. 2019. *SME and Entrepreneurship Outlook 2019*. Paris: OECD publishing.

Piore, M., and C. Sabel. 1985. *The Second Industrial Divide: Possibilities for Prosperity*. New York: Basic Books.

Poletaev, M., and C. Robinson. 2008. "Human Capital Specificity: Evidence from the Dictionary of Occupational Titles and Displaced Worker Surveys, 1984–2000." *Journal of Labor Economics* 26: 387–420. doi:10.1086/588180

Rigby, D. 2015. "Technological Relatedness and Knowledge Space: Entry and Exit of US Cities from Patent Classes." *Regional Studies* 49 (11): 1922–1937. doi:10.1080/00343404.2013.854878

Vona, F. 2021. *Labour Markets and the Green Transition: A Practitioner's Guide to the Task-Based Approach.* Edited by F. Biagi, and A. Bitat. Luxembourg: Publications Office of the European Union.

Wixe, S., and M. Andersson. 2017. "Which Types of Relatedness Matter in Regional Growth? Industry, Occupation and Education." *Regional Studies* 51 (4): 523–536. doi:10.1080/00343404.2015.1112369

World Economic Forum. 2020. *The Future of Jobs Report 2020.* Geneva: The World Economic Forum.

Living innovation machines: modelling innovation in time and space variable-geometry territorial units using machine learning

Rafael Boix-Domenech, Vittorio Galletto, Fabio Sforzi and Francesco Capone

ABSTRACT
This paper investigates the innovative intensity (iMID effect) in the local production systems (LPSs) using dynamic territorial units that modifies their geographical boundaries and specialization over time. The paper is based on the idea that the local communities are the real unit of generation of innovation and that they are constantly changing, behaving like a 'living innovation machine'. To explore this dynamic relationship, the paper applies the Marshallian industrial district (MID) as an approach to industrial change. Then, it focuses on the Marshallian industrial districts (MIDs) as a type of LPS and asks how the innovation effect changes in MIDs and other types of LPSs when dynamic territorial units are used. The differences in innovation intensity between the different types of LPSs are explained for Spain in the period 1991–2014 using new methods of analysis that combine causal analysis, a variable and adaptive geometry of territorial units and industrial specialization, and machine learning methods. The results show that the transformation of an LPS into another type of LPS does not automatically imply a change in innovative capacity, and the type of LPS of origin continues to be relevant in explaining current innovation processes.

1. Introduction

In the canonical definition of Giacomo Becattini (2004a), a Marshallian industrial district (MID) is a community of people and a segmented population of firms – i.e. specialized in products, parts of a product or different phases of the same production process – that share a historically and geographically determined place, giving rise to a local community.

The basic aim of the MID is to promote the welfare of one's community of people. To achieve this, what matters is not the production of goods, but the ability to generate innovations. From this point of view, the MID can be seen as a social system of knowledge production.

Despite a certain belief in the literature that the MID is a producer in the mature part of the product life cycle, the reality, quite the opposite, is that it is a 'modern manufacturer'. The characteristic task of the MID is 'that of creating new wants by showing people something which they had never thought of having before; but which they want to have as soon as the notion is suggested to them' (Marshall 1920, 234).

The continuous production of innovation requires a constant change and adaptation of the MID, so that it can be conceived as a 'living innovation machine'.

The intense and continuous generation of innovations is the origin of the so-called iMID effect (innovation-district effect) introduced by Boix and Galletto (2009). The iMID effect defines the fact that MIDs as a model of production exhibit higher innovative intensity than the rest of the national economy.

In the view of Boix and Galletto (2009), the iMID effect is not restricted to the MID as a model of production, but it extends to all types of local production systems (LPSs) that make up the nation's economy. As such, it allows a comparison of the innovative capacity among MIDs and other types of LPSs. This view is consistent with Becattini's view of the MID as an approach to industrial change (Becattini 2004a).

Strong evidence of the iMID effect has been found in the Spanish and Italian economies, where the MIDs innovate between 24% and 48% more than the average of LPSs (Boix, Galletto, and Sforzi 2019; Boix, Capone, and Galletto, 2022).

We can differentiate two generations of works on the iMID effect. The first generation measures the iMID effect at a single point in time (Boix and Galletto 2009; Boix and Trullén 2010; Galletto and Boix-Domènech 2014). The second generation considerably broadens the temporal coverage of the data, confirming the iMID effect over a long period (Boix, Galletto, and Sforzi 2019; Boix, Capone, and Galletto, 2022).

This second generation uses an operational simplification: it fixes the territorial pattern of LPSs at a point in time and considers that they are constant throughout the analysed period. However, we know that this is not the case. Sforzi and Boix (2019) show how in Italy and Spain the number of LPSs changes over time redefining both their boundaries and their type of industrial specialization. Between 1991 and 2011, the number of LPSs fell from 784 to 596 in Italy and from 704 to 483 in Spain. Consequently, the number of MIDs also reduced from 199 to 120 in Italy and from 210 to 151 in Spain (Sforzi and Boix 2019).

What is relevant in this process is what lies behind these numbers. In 2011, there are not only fewer MIDs than in 1991, but they are also different. First, they are bigger because they have expanded their geographical boundaries or merged with other LPSs. Second, there have been transitions: some MIDs turned into other types of LPS, and some LPSs that were not MIDs became MIDs. These changes may question the validity of the results obtained in previous studies on the measurement of the iMID effect and suggest the use of dynamic territorial units as a research priority.

It is on this crucial point that the main contribution of this paper is focused. It investigates how the innovation effect changes in the LPSs using dynamic territorial units that modify their geographical boundaries and type of industrial specialization over time. In the framework of the literature on MIDs, the paper maintains the idea that LPSs are local communities that generate innovation and that are constantly changing, behaving like a living innovation machine. Consequently, the analysis of innovation processes in time, based on static territorial units, can be flawed and is acceptable only in the preliminary stages of research.

Another fundamental problem, related to dynamic territorial units, is how to measure the causal effect of being a MID on innovative intensity. The comparison between the average innovative intensity of the MIDs and the average of the national economy tells us that the MIDs are more innovative, but it does not tell us if all this difference is because they are MIDs or whether this is due to other causes. For this reason, Boix and Galletto (2009) introduced in their initial research a causal approach, using a place-based function of production of innovations. The continuous redefinition of the territorial units of analysis and the dynamics of transitions between types of LPSs requires rethinking both the causal framework of the innovation production function and the use of measurement methods that are more flexible than those used in previous works (Boix and Galletto 2009; Boix, Galletto, and Sforzi 2019).

The article raises one question: how does the innovation effect change when dynamic territorial units are introduced? To answer this question, the paper focuses on the change of the iMID effect in the LPSs of Spain during the period 1991–2014. Altogether, we can summarize the novelties of the paper into three points. First, it is one of the few research projects on the geography of innovation that uses a variable and adaptive geometry of territorial units and LPSs that change their type of industrial specialization over time, closing a relevant gap in the empirical literature. Second, the paper proposes a methodology that allows the changes in the innovative intensity of the MIDs and other LPSs to be explained according to their geographical and typological transformations. Third, big data and new machine learning methods are used as a more flexible way to explain the change of the innovative intensity in the LPSs.

The paper is organized as follows. The next section provides a brief discussion to frame the change of the territorial systems and MIDs. The third section redefines the causal framework and describes the methodology and the data used. The fourth section describes the data and the dynamics of transition of the LPSs in Spain. The fifth section explains the machine learning methods for measurement of the innovation effect. The sixth section provides the causal measurement of the iMID effect. The seventh section concludes and highlights implications for future research.

2. MIDs as living innovation machines

The historical antecedents of MIDs date back to Alfred Marshall's *Principles of Economics* (1920), and its functioning is inexorably linked to those of the Marshallian producer and consumer and his/her search for variety. In a Marshallian system, there is a close relationship between the emergence of consumer desires and the way that producers satisfy them. The so-called 'Marshall consumer' (Swann 2009, 34) is an innovator because he/she is always seeking to satisfy a desire for variety and change, and social distinction. The Marshall's producer creates these needs, because he/she has the ability of 'forecasting the broad movements of production and consumption, of seeing where there is an opportunity for supplying a new commodity that will meet a real want or improving the plan of producing an old commodity' (Marshall 1920, 298). A MID is the most suitable production machine for satisfying the Marshall consumer. The goods that MIDs produce are innovative and customized, being produced in response to a differentiated and variable demand (Becattini 1990).

The MID resolves the tension between the implementation of routines on a large scale (not that of the individual firm, but that of the population of firms) and the initiative and variety which provides many firms operating on a small scale, coordinated with each other. The large firms improve efficiency by scaling up routines, although this can reduce flexibility and creativity, and a smaller total number of firms reduces variety, which in turn reduces welfare (Loasby 2002). The MID ensures 'the "automatic organisation" ... of highly specialized activities while facilitating both the generation and the active discussion of novel ideas, including ideas for constructing new patterns of relationships between firms' (Loasby 2002, 667).

As a local innovation-producing community, the MID suggests that potentially all LPSs, which are equally local communities although structured differently from the MID as regards the key features of the community of people and the population of firms, can enjoy a positive or negative innovation effect similar to the iMID effect; although the underlying causes of the iMID effect will be different to the extent that a generic LPS is different from a MID. The main consequence of this statement is that each LPS can be described as a living innovation machine. As such, LPSs change over time and space, as they constantly redefine their internal socio-economic structure as well as their external geographical boundaries.

The literature on change of territorial systems in economic geography and in regional and urban economics converges on this problem. We find it in Hägerstrand's time–space geography (Hägerstrand 1970), in the modifiable areal unit problem and the geographical analysis machine (Openshaw and Taylor 1981; Openshaw, Charlton, and Craft 1988), in changes in specialization models and variable geometry territorial units (Trullén, Boix, and Galletto 2013) and in the redefinition of policies for complex systems (Cooke 2013).

The MID is an organic unit subject to constant change. Let's take some examples: the transformation of the 'urbanized countryside' of Tuscany, the evolution of Prato (Becattini 2004b) and the districtualization processes (Lazzeretti and Capone 2017), or the possible death of MIDs in the face of globalization (Dei Ottati 2018). The identification of MIDs in Italy and Spain has been repeatedly updated to reflect their changing dynamics over time (Sforzi and Boix 2019).

From the previous literature, we can synthesize that the positive or negative changes in the innovative intensity of an LPS could be given by three stylized situations. First, the change can be the consequence of a purely endogenous improvement effect without the LPS changing its morphology (i.e. the boundaries) or its typology (i.e. the model of production or the key industry). Second, the change can be due to the fact that the LPS has changed its morphology, and this alters the value regardless of whether there is endogenous improvement. The morphology can change due to the expansion or contraction of the boundaries of the LPS (e.g. incorporating or disjoining nearby localities), or by processes involving the fusion, division or absorption of other LPSs. And third, a LPS can transform into another type of LPS, a situation that could have implications for its capacity and way of innovating. The typological transformation can be due to the endogenous change of the LPS, but also to the exogenous change due to the fusion, division or absorption of other LPSs. Empirical evidence suggests that changes in innovative capacity are not pure, but are combinations of some of the above situations.

3. Living innovation machines and innovation intensity: a matter of causality

3.1. The place-based knowledge production function

The measurement of the iMID effect has traditionally been carried out in two ways. The first, makes a direct comparison of the innovative intensity of the MIDs against other types of LPSs, using indicators such as the mean or the median (Boix, Capone, and Galletto 2022). The second, through econometric modelling based on an innovation production function, brings the measured effect closer to a causal effect.

The innovation production function used by Boix and Galletto (2009) consists of two parts. The first part relates the production of innovations to research and development (R&D) expenditures (Griliches 1992). In the second part, the typology of LPSs is added, turning it into a place-based innovation production function (Boix, Galletto, and Sforzi 2019):

$$i_{t,j} = \gamma r^{\beta}_{t-1,j} \delta_{t,k} \varepsilon \qquad (1)$$

where i is the average innovation per employee; r is the average R&D expenditures per employee; t refers to the time period; j refers to the LPS; k refers to the type of LPS; γ, β and δ are parameters and ε is a nuisance.

3.2. Causal schemes and backdoors in the knowledge production function

The innovation production function has some limitations. To overcome them, we have transferred the mechanism to a directed acyclic graph (DAG) (Pearl and Mackenzie 2018).

In Boix and Galletto (2009), it is implicitly assumed that the type of LPS does not determine the R&D expenditures. Under this assumption, the estimated δ coefficients correspond to the total effect of each type of LPS on the innovative intensity, with all this effect being direct (without mediators). However, it is more realistic to assume that the types of LPS determine the R&D expenditures, or at least the private R&D expenditures. In this case, the 'total effect' on the innovative intensity cannot be estimated due to adjustment for intermediate covariates such as the private and public R&D expenditures per person employed. What Boix and Galletto (2009) are estimating is the 'direct effect' of each type of LPS on the innovative intensity, 'not the total effect'. A part of the effect, the indirect one, would not be accounted for.

In Boix, Galletto, and Sforzi (2019), private and public R&D expenditures are lagged one period because transforming R&D into innovation takes time (Griliches 1992). Again, the authors assume that past R&D expenditures do not depend on the current type of LPSs, which in this case is logical. However, it is reasonable to assume that past R&D expenditures depend on the past type of LPSs, which would lead to a causal scheme like the one in Figure 1. In this scheme, three backdoors are open, connecting innovative intensity with the current type of LPSs, and going through the two variables of R&D expenditures and the past type of LPSs. There are two ways to block these backdoors: controlling for R&D expenditures or controlling for the past type of LPSs. Boix, Galletto and Sforzi (2019) opt for the first, thus estimating this time the total causal effect of the typology of LPSs on innovative intensity.

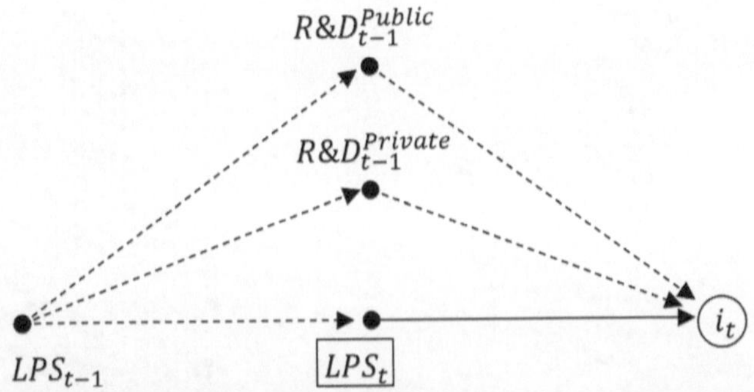

Figure 1. Basic relationships in the directed acyclic graph*. * The dashed lines involve potential biasing paths that open backdoors. The superindex *t* indicates the period, where t = current period and t-1 = one lagged period.

The alternative in Figure 1 is to use the past type of LPSs as a control variable. In this case, we should not include the R&D expenditures because it opens a backdoor between the innovative intensity and the past type of LPS. This alternative has the advantage that it can be applied to situations where it is difficult to obtain R&D expenditures or R&D measurement is not precise.

3.3. One more step: controlling the role of policy, inertia in the innovation process, physical and human capital and agglomeration economies

Some criticism of previous work on the iMID effect stems from the very simple structure of the innovation production function. This section integrates into the causal scheme other important factors that can affect the innovative intensity of LPSs. There are precedents for some of them, such as human capital and agglomeration economies. The rest are introduced for the first time here: the role of policy, the effect of past innovative intensity on the present one (which can lead to inertia in innovation), and physical capital. Figure 2 separates these variables into partial DAGs to facilitate their interpretation. The result is the same if we integrate them all into a single DAG.

3.3.1. Economic and innovation policy

The DAG in Figure 2a introduces endogenous economic or innovation policy that depends on each LPS, and exogenous economic policy that depends on institutions such as the regional or national governments. Private spending on R&D is now determined by the decisions of local firms and by endogenous and exogenous economic policy. Public spending on R&D is determined by endogenous and exogenous economic policy. There are now two possible interpretations:

1. We are interested only in the effects of the current type of LPS on innovative intensity. For practical purposes, as economic policy is mediated by R&D expenditures, it is only necessary to condition the estimates by the R&D expenditures or by the past type of LPS, as in the previous case, which avoids the problem of obtaining variables

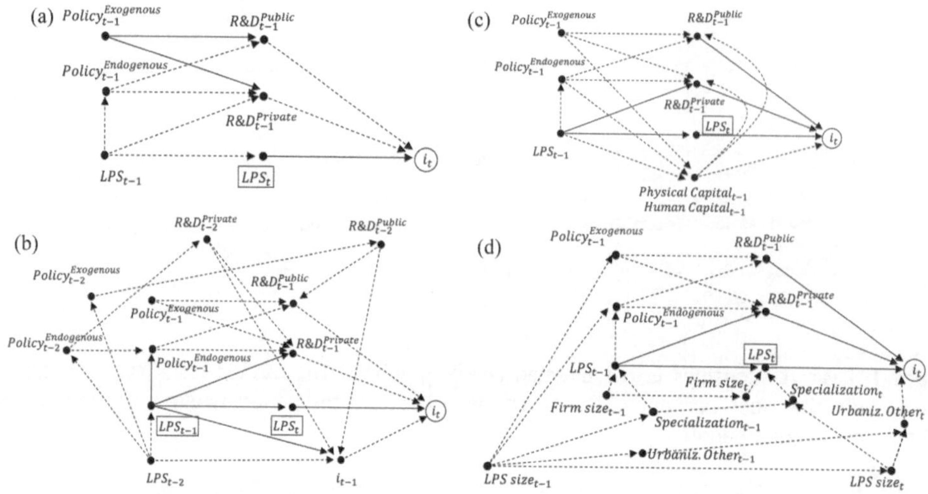

Figure 2. DAGs with the effects of past types of LPSs and additional controls (economic policy, capital intensity, productive and organizational structure, and agglomeration economies)*. *The dashed lines involve potential biasing paths that open backdoors. The superindex *t* indicates the period, where *t* = current period, *t*-1 = one lagged period, and *t*-2 = two lagged periods.

that approximate economic policy decisions. It could also be conditioned by endogenous and exogenous economic policy plus private R&D expenditures.

2. We are interested in the causal effects of the current but also of the past type of LPS on innovative intensity. In this case, the only covariates that we should include in the estimation are the current and past type of LPS. If we include the R&D variables or the policy variables, we open backdoors between the innovative intensity and the typology of LPS, and we would end up estimating the direct effect instead of the total effect.

3.3.2. Physical and human capital

Figure 2b introduces the stock of physical capital per worker and human capital (education, training or learning), which are partly determined by economic policy. We assume that the effects of physical and human capital are also transmitted with a certain delay to the generation of innovations (the result for the causal graph does not change if it is considered that they are instantaneous) and that they also condition R&D expenditures. Again, we have two situations: in the first, we are interested in the effects of the current type of LPS on the innovative intensity. In this case, to close all the backdoors, it is enough to control by the past type of LPS, without including any other variable. In the second situation, our interest is the causal effect of the current and past type of LPS. In this case, it is not necessary to include more variables than the current and past types themselves. In fact, physical and human capital are colliders, and controlling for them would open backdoors and bias the estimates (Pearl and Mackenzie 2018).

3.3.3. Inertia in innovation

We can also consider the situation in which there is inertia in the generation of innovation; that is, past innovative intensity affects current innovative intensity (Figure 2c). Following the circuit of the other DAGs, in which the effect of R&D expenditures acted with a delay, it is considered that past innovative intensity is determined by the previous type of the LPSs lagged one and two periods (at *t-1* and *t-2*). There is a temptation to solve the problem by controlling for the lagged innovative intensity (i^{t-1}). However, the lagged innovative intensity is endogenous and is a descendant of the type of LPS at *t-1*. Including an innovation delay would only make things worse and we would need an instrument to do it correctly. Fortunately, all the backdoors pass through the type of LPS at *t-2*, so it is enough to control for *t-2* to close them and to be able to estimate the total causal effect of the current and *t-1* type of LPS. Indeed, if the only causal effect we are interested in is that of the current type of LPS (not the past one), controlling for the past type of LPS would suffice to close all the backdoors.

3.3.4. Productive structure and agglomeration economies

Finally, we include environment variables related to the productive structure and agglomeration economies (Figure 2). We differentiate between internal economies and localization economies, on the one hand, and urbanization economies on the other. Internal and localization economies should not be included in the estimates because the type of LPS is determined empirically using data on firm size and industrial specialization in the LPS (Sforzi and Boix 2019). Their inclusion in the estimates absorbs the iMID effect and biases the results (Boix and Galletto 2009).

Figure 2d introduces the role of urbanization economies, such as the population size and other factors such as variety or population density. Variety and density could affect innovative intensity directly or indirectly (via the type of LPS). In the first case, backdoors are generated. The easiest way to close all the backdoors is to include the current and lagged size of the LPS, for example using the population or the number of persons employed.[1]

3.3.5. Synthetic causal forms

Table 1 synthesizes the causal forms that arise from the previous DAGs (Figures 1 and 2) and the effects that are controlled in each of them. These synthetic causal forms allow estimating the causal effects of the current and past type of LPSs on the innovative intensity including a relatively reduced number of variables. At the same time, these causal forms make it possible to control for a large number of effects without the need to

Table 1. Summary of causal effects of the type of LPS on the innovative intensity.

Model	Target	Causal form	Features controlled by the causal form				
			R&D	Economic and innovation policy	Physical and human capital	Inertia in innovation	Productive structure and agglomeration economies
1	LPS_t	$i_(t,j) = f(r'_{(t-1,j)}\ \delta_(t,k))$	x	x			
2	LPS_t	$i_{t,j} = f(\delta_{t,k}, \delta_{t-1,k})$	x	x	x		
3	LPS_t, LPS_{t-1}	$i_{t,j} = f(\delta_{t,k}, \delta_{t-1,k}, \delta_{t-2,k})$	x	x	x	x	
4	LPS_t, LPS_{t-1}	$i_{t,j} = f(\delta_{t,k}, \delta_{t-1,k}, \delta_{t-2,k}, LPS\ size_t, LPS\ size_{t-1})$	x	x	x	x	x

t = current period; *t-1* = period lagged one time; *t-2* = period lagged two times; *i* = innovative intensity; *r* = R&D expenditures; δ_k = LPS of type *k*.

include in the estimates variables which are difficult to measure, such as the economic policy, the local stock of physical capital, etc.

4. Data

4.1. Data and indicators

The measurement of technological innovation follows previous research on the iMID effect (Boix and Galletto 2009; Boix, Galletto, and Sforzi 2019) and is based on instruments for the protection of intellectual property related to technology, such as patents and utility models. We have used the innovation database of Boix, Galletto, and Sforzi (2019), which allows comparison with the results of their work which uses static types of LPSs centred on 2001.

For the whole period 1991–2014, the database includes more than 143,000 patents and utility models by application year, without previous priority claims to avoid duplication, and geo-referenced first by municipality and then by LPS based on inventors' address. In case of multiple inventors, a fractional assignation is made to the different municipalities (national and foreign) of the addresses. The sources of data are the Spanish Patent and Trademark Office (OEPM), European Patents (EPO – PATSTAT) and U.S. Patents (USPTO – PatentsView). The number of patents and utility model is divided by the number of employees of the LPS to obtain the innovative intensity.

R&D expenditures are divided into private (firms) and public (university and other public research entities).[2] Unlike previous works which use the yearly flows of expenditures (e.g. Boix and Galletto 2009; Boix, Galletto, and Sforzi 2019), we use the stock of R&D in the previous period. This measurement considers the accumulation of the expenditures and the delay between the expenditures and the generation of the innovations (Griliches 1992). To calculate the stock, we used a yearly depreciation of 20% of the value of each year, implying that after five years (the legal amortization period of R&D in Spanish business accounting) the value of the R&D is null. The stock is divided by the number of employees in the LPS.

The number of employees for all the indicators is obtained from the Social Security registers of the Spanish Ministry of Employment and includes wage-earning employees, self-employees and special regimes.

To avoid yearly fluctuations and lags in the outcome of innovation processes (Griliches 1992), the data have been aggregated in six periods of four years (1991–1994, 1995–1998, 1999–2002, 2003–2006, 2007–2010, 2011–2014). As in Boix, Galletto, and Sforzi (2019), the selection of the period is marked by the availability of maps of MIDs (Sforzi and Boix 2019).

4.2. The dynamic territorial unit of analysis and its change

The MID approach to industrial change states that change is a socio-economic process that takes place neither in a firm nor in an industry, but within a local community due to the way in which firms, families and institutions interpenetrate (Becattini 2004a; Sforzi and Boix 2015). As a consequence, the appropriate territorial unit to measure industrial change and innovation is the local community. In the MID literature, the local community has been empirically approximated through the definition of the local labour system (LLS) (Sforzi 2012).

The boundaries of LLSs change over time, due to the territorial mobility of people and firms, as job or profit opportunities change. And of course, the various external contexts, institutional and economic, play their role, now creating constraints, now opportunities. The change of boundaries, as well as their stability, can also lead to a change in the industrial characteristics of the LLSs. When LLSs are regarded from this perspective, the label of local production system (LPS) can be applied. LPSs define different industrial types, one of which matches the concept of the MID as a model of production (Sforzi and Boix 2019).

The identification of LPSs and their categorization in MIDs and non-MIDs have been carried out for Spain in 1991, 2001 and 2011 by Sforzi and Boix (2019) using the Sforzi-ISTAT algorithm (Sforzi 2012). Following Boix, Galletto, and Sforzi (2019), non-MIDs have been classified into five types based on the first step of the Sforzi-ISTAT algorithm: manufacturing LPSs of large firms, LPSs specialized in agriculture and mining, LPSs specialized in construction, LPSs specialized in business services and LPSs specialized in other services (including consumer, social and traditional services) (Table 2).[3]

Since the innovation data is grouped into four-year periods to reduce its variability, the classification of LPSs closest to the time point of identification of the LPSs is used. The result is six four-year periods, where the type of LPSs changes every two periods. In this way, we begin by applying the type of LPS in 1991 to the 1991–1994 period, and end by applying the type of LPS in 2011 to the 2011–2014 period (Table 2).

Between the first and last period, the number of MIDs is reduced from 210 to 159, although their share of the total number of LPSs increases slightly. The composition of the MIDs has also changed from 1991 to 2011. In terms of typology, some MIDs have transformed into other types of LPSs and vice versa. The most notable is the merging and transformation of agricultural LPSs into MIDs (Figure 3), which partly explains the phenomenon of agri-food industry in Spain and its emergence as the most important industry within Spanish manufacturing.

The transformation of MIDs into manufacturing LPSs of large firms is also notable. Figure 3 shows an intense process of transformation of the LPSs in all directions, which highlights the idea that the productive profiles of local communities are constantly changing.

Table 2. LPSs by type, 1991, 2001 and 2011.

Year of identification	1991		2001		2011	
Period number	1 and 2		3 and 4		5 and 6	
Applied to the periods	1991–1994, 1995–1998		1999–2002, 2003–2006		2007–2010, 2011–2014	
LPSs by type	Total	%	Total	%	Total	%
Agriculture and mining	384	54.5	276	40.8	163	33.0
Construction	30	4.3	42	6.2	43	8.7
MIDs	210	29.8	215	31.8	159	32.2
Manufacturing of large firms	16	2.3	56	8.3	59	11.9
Business services	0	0	3	0.4	2	0.4
Other services	64	9.1	85	12.6	68	13.8
Total	704	100.0	677	100.0	494	100.0

Source: Sforzi and Boix (2019), Boix et al. (2018) and own elaboration.

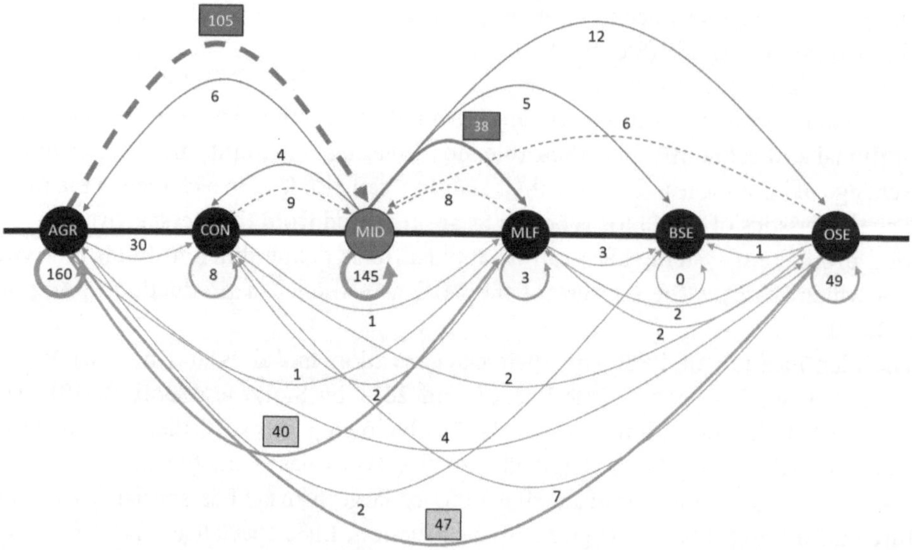

Figure 3. Arc plot with the total number of transformations between types of LPSs from 1991 to 2011. The figures represent the number of LPSs involved in the transformation. Type of LPS: AGR = Agricultural and mining; CON = Construction; MID = Marshallian industrial districts; MLF = Manufacturing LPSs of large firms; BSE = Business services LPSs; OSE = Other services LPSs. Transformations include endogenous evolution or merging/division in another LPS with another dominant specialization.

5. Machine learning methods for the measurement of the iMID effect: decision trees and random forests

The measurement of the conditioned iMID effect poses some problems, mainly derived from the high number of LPSs that do not innovate in some of the periods (around 25%) and that the typology of LPS is a categorical variable that can take six values. Boix and Galletto (2009), Boix and Trullén (2010) and Galletto and Boix-Domènech (2014) use an innovation production function linearized by taking logarithms. Since the logarithms of the LPSs with zero innovations cannot be computed, the authors treat the problem as a sample selection process and use the Heckman estimator for panel data. Boix, Galletto, and Sforzi (2019) estimate a nonlinearized production function using a quantile model with nonadditive fixed effects. Both solutions give reasonable results, although they pose limitations derived from the assumptions of the selection models, the role that the fixed effects play in the method, the consistency of the fixed effect estimators and the heterogeneity of the innovation per employee across the LPSs.

In this paper, we use new Machine Learning (ML) methods. ML has begun to be applied recently to the resolution of problems in social sciences by leading scientists such as Hal Varian (Varian 2014) or Nobel laureate Guido Imbens (Athey and Imbens 2019). Overall, the ML approach gives us three advantages: it does not need to impose any structure on the data, it makes available a huge variety of flexible methods, and it enables new ways to analyse and communicate the results.

Among the new ML methods there are two that, while being simple and well known, can be particularly useful for estimating the iMID effect and providing complementary outputs: decision trees and random forest.

Decision trees make predictions and generate interpretable results by splitting the instances (the LPSs by period in our case) into binary branches, asking in each branch which is the variable that best predicts the data and from which value of the variable the best prediction is produced. We implement Conditional Inference Trees (CITree) (Hothorn, Hornik, and Zeileis 2006), which use permutation tests to make the divisions of the nodes and are more robust than other tree methods to overfitting and variable selection bias. CITree presents the typical advantages of decision trees: they are nonparametric, manage complex relationships between features and interaction between variables, are robust to outliers and insensitive to monotone transformations, and they are not affected by unimportant features or by cross-relationships between features that generate multicollinearity. They do not produce a single coefficient like traditional regression methods, but instead supplies sets of rules, which allows better targeting of policy strategies.

Decision trees are sensitive to sample variation and can show a trade-off between bias and variance. One way to solve this problem is using random forests. A random forest (Breiman 2001) is an ensemble of trees created by adding randomness to the sample and variables. The method bootstraps the original sample to create many samples and then creates a decision tree for each bootstrapped sample using in every split only a random sample of the predictors. At the end of the process, for continuous features we average the predictions by instance, and for categorical features we use the classes with more votes. By introducing randomness in both the instances (observations) and the features (variables), the variance is reduced without the trees being directly correlated with each other and the bias does not increase. The predictive capacity is notably increased and many of the inconveniences of single decision trees are solved. The method is relatively robust to changes in bootstrapping or the number of variables that can be used in the splits.

The main disadvantage of random forests regarding the single trees is that the structure that results from the ensemble is so complex that it is very difficult to interpret directly. We use an Accumulated Local Effects (ALE) plot to explain the effect of the different types of LPSs on the average prediction of the random forest. ALE values (Apley and Zhu 2020; Molnar 2022) show the relative effect of changes in the feature on the prediction conditional to a given value (Molnar 2022) and centred on the mean prediction. They are unbiased even when the features are correlated.

In the random forest the typology of LPSs is introduced as a set of dummy variables (one-hot encoding), and without the need to omit any since the forest is not sensitive to collinearity. The ALE of a dummy only has two values (0 and 1) so that we can obtain the conditional marginal effect of a type of LPS as the difference between the ALE values at points 0 and 1: MgE = ALE (LPS = 1) – ALE (LPS = 0). Likewise, we can obtain the relative effect by dividing this effect by the prediction of the outcome variable: RMgE = MgE/mean($\hat{y}$).

6. Results

6.1. The long-term change of the innovative intensity in the LPSs

Figure 4 shows the average innovative intensity of the different types of LPSs between 1991 and 2014. The territorial pattern of LPSs changes every two periods, which

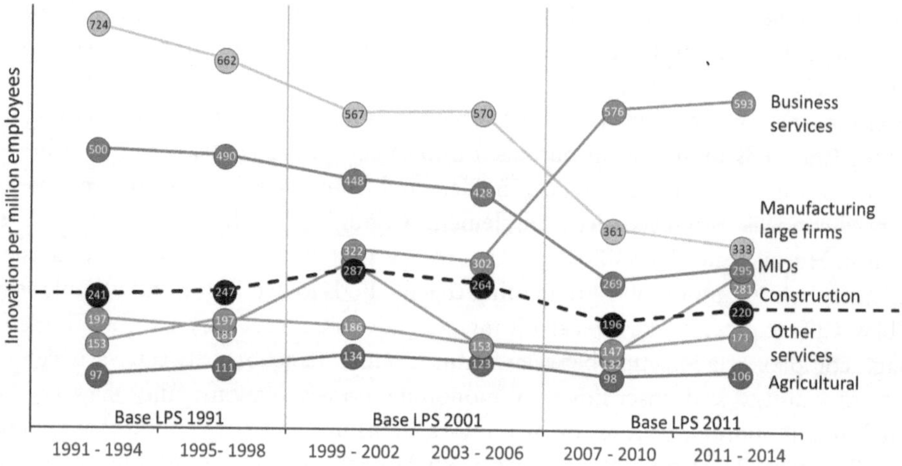

Figure 4. Innovation per million employees per type of LPS and period. Average innovative intensity by type of LPS. Source: our elaboration from Boix et al. (2018), Sforzi and Boix (2019), Census (INE), Ministry of Employment, OEPM, EPO, and USPTO.

makes it possible to follow the innovative intensity without assuming that the territorial unit of analysis is static. The MIDs show an innovative intensity superior to the average of the rest of LPSs during the whole period. The difference is greater between 1991 and 1998, when the innovative intensity of the MIDs reaches twice the average of LPSs. It reduces to around 34%–37% in the last periods (2007–2010 and 2011–2014), coinciding with the 2008 crisis.

6.2. Causal estimates of the iMID effect

The ML methods employed are robust to overfitting and can be used without cross-validation. The suitability of not using cross-validation has been verified by dividing the sample into a part of training (periods 2, 3 and 5) and test data (periods 4 and 6). The percentage of explained variance is almost identical in both samples (around 0.14 and 0.13 in the tree and 0.18–0.16 in the forest) and the structure of the CITree does not change. Consequently, the complete sample of periods 2–6 has been used for the estimations (period 1 is not included to be able to use the variables lagged one period).

Figure 5 shows the output of the CITree for model 1 during the five periods from 1995 to 2014. The average innovative intensity of the sample and of the prediction is 247 patents per million of persons employed. The greatest innovative intensity is obtained in the right branches of the tree, whose first characteristic is that most of the LPSs are MIDs, manufacturing LPSs of large firms, or business services LPSs. Those with R&D expenditures above € 564 show an average innovative intensity of 2.4 times average intensity, and those below € 564 show an innovative intensity 30% higher than average (15% and 25% of the instances, respectively). The left branches of the tree correspond to the rest of types of LPSs (agriculture, construction, other services) and show an innovative intensity mostly below the average.

The results structure of the tree is interesting although for our problem it lacks precision, as it fails to distinguish the innovation effect for each individual type of LPS

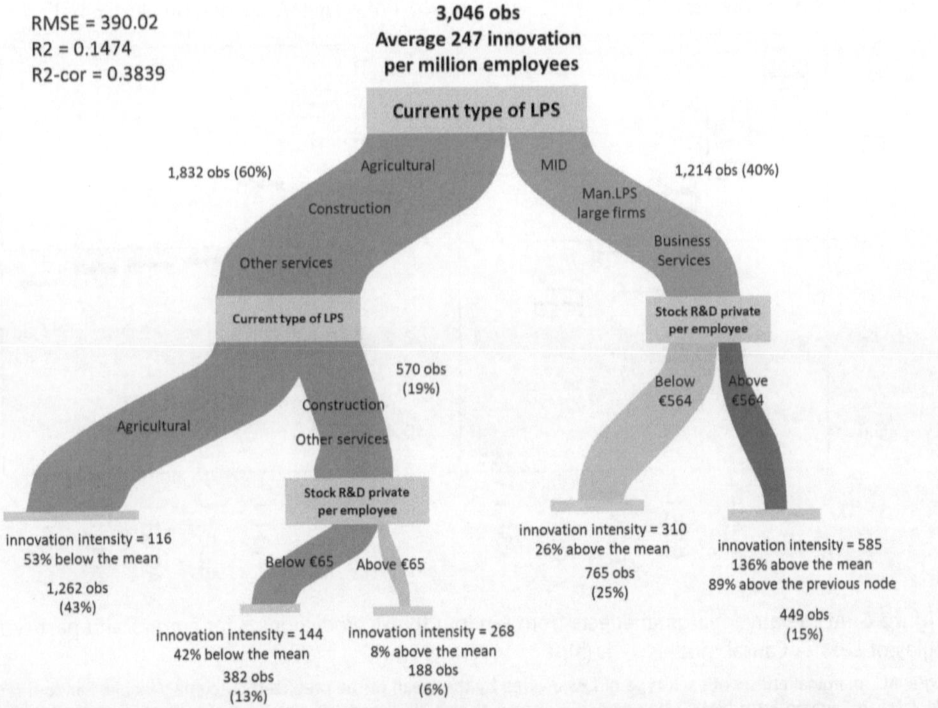

Figure 5. Conditional inference regression trees for Model 1: $i_{t,j} = f(\gamma r_{t-1,j}, \delta_{t,k})$. iMID effect in the current period under R&D, policy, and physical and human capital.

and tends to group some of them together. We have found this same problem to directly individualize the effects by type of LPS for all the models in Table 1. As the objective is to obtain separate estimates of the effects for each type of LPS, for the rest of the models we will provide only the results of the random forest.

Figure 6 summarizes random forest estimates for the typology of LPS for the causal models of Table 1 (detailed results can be found in Annex 2). These models control for everything from the effects of R&D to the effects of agglomeration economies. The relative marginal effects for each type of LPS are calculated from the ALE values (see Section 5) and can be interpreted in a similar way to the elasticities of other works on the iMID effect (Boix and Galletto 2009; Boix, Galletto, and Sforzi 2019).

The differential innovative intensity (iMID effect) of being MID in the current period varies between a maximum of 39% (model 1) and a minimum of around 13% (models 3 and 4) (Figure 6 and Annex 2). The iMID effect of model 1 is very similar to the 35% estimated by Boix, Galletto, and Sforzi (2019) using the same theoretical model, although in their case with R&D flows instead of stocks and fixing the LPSs at a single point in time (static territorial unit of analysis). The iMID effect is reduced by half (19.5%) when controlling for inertia in the innovation process (model 2), which indicates strong inertia in innovation in MIDs. As in previous research, the innovative intensity of the manufacturing LPSs of large firms is slightly higher than that of the MIDs.

Another advantage of using dynamic types of LPSs is that for the first time we can measure the effect of the past type of the LPS on current innovative intensity. The

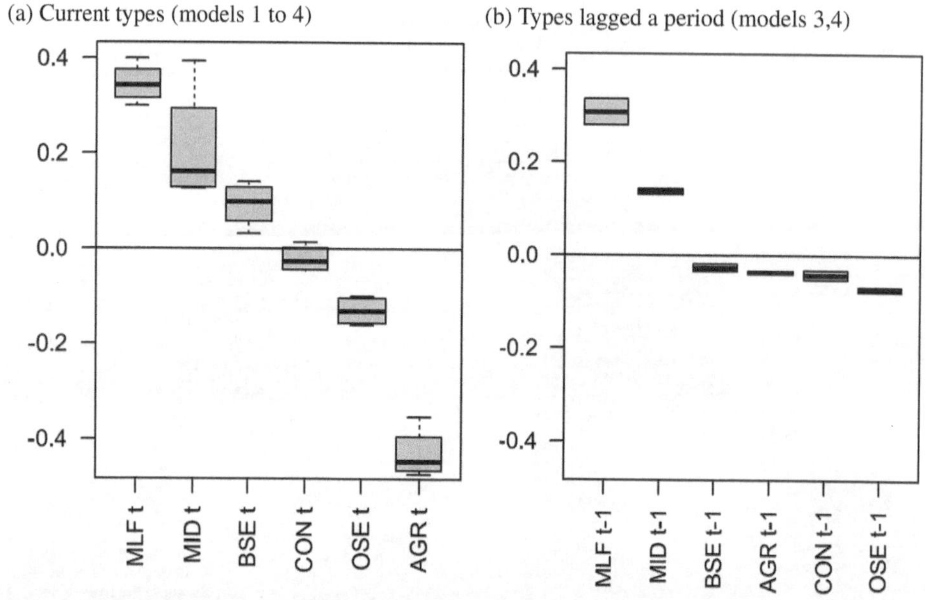

Figure 6. ALE relative marginal effects from random forest predictions[a] for current and past typologies of LPSs[b]. Causal models 1–4. (a).

Note: ALE marginal effects of each type of LPS divided by the mean of the predicted outcome variable. Random forest Root Mean Squared Error (RMSE) between 366 (model 4) and 396 (model 2). (b) Type of LPS: AGR = Agricultural and mining; CON = Construction; MID = Marshallian industrial districts; MLF = Manufacturing LPSs of large firms; BSE = Business services LPSs; OSE = Other services LPSs.

effect of being MID in a previous period (models 3 and 4), ranges between 13 and 14%, that is, similar to that of being MID in the current period (Figure 6b). If we add the current and past iMID effect, the cumulative effect would be around 26%.

We can now go one step further, combining the current and past type of the LPSs in a kind of transformation matrix like the one suggested in Figure 3. In this case we would have a new set of dummies where, for example: ($MID_{t-1} \rightarrow MID_t = 1$) indicates that the LPS was MID in the previous period and continues to be in the current period; ($AGR_{t-1} \rightarrow MID_t$) indicates that the LPS has transitioned from agricultural LPS to MID, and so on. The random forest allows us to introduce all these relationships without the regularization or collinearity problems of traditional models and obtain the marginal effects from the ALE values.

Figure 7a shows the detail of the effects of the transformations involving the MIDs. The LPSs that were MIDs in the previous period and continue to be MIDs in the current period show an innovative effect of between 27% and 30% higher than the average of LPSs. The manufacturing LPSs of large firms that have become MIDs show an innovative intensity of between 34% and 42% higher than the average of LPSs (models 4 and 3, respectively). The effect is lower for other transformations, and clearly negative for three of them, especially for the agricultural LPSs that transform into MIDs. These results suggest that typological transformations do not automatically imply a change in innovative intensity, but rather that in many cases this takes place more slowly and that the starting point (past type of LPS) influences the result.

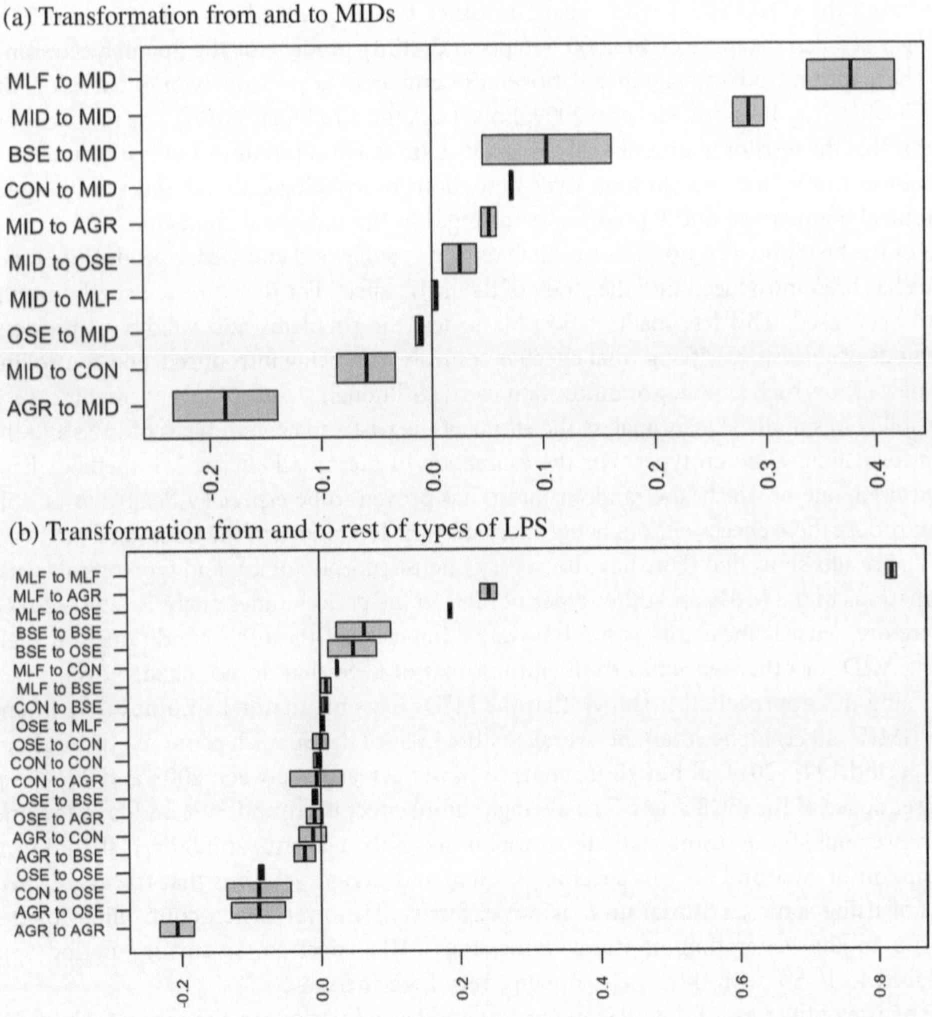

Figure 7. ALE relative marginal effects from random forest predictions[(a)] for transformations between types of LPSs[(b)]. Causal models 3 and 4.

Note: (a) ALE marginal effects of each type of LPS divided by the mean of the predicted outcome variable. Random forest Root Mean Squared Error (RMSE) between 367 (model 4) and 373 (model 5). (b) Type of LPS: AGR = Agricultural and extraction; CON = Construction; MID = Marshallian industrial districts; MLF = Manufacturing LPSs of large firms; BSE = Business services LPSs; OSE = Other services LPSs.

Figure 7b shows the results for the rest of the transformations. Clearly, those involving manufacturing LPSs of large firms show the greatest effect, and in particular those LPSs that maintain their specialization as manufacturing LPSs of large firms, whose effect exceeds 80%.

7. Conclusions

The iMID effect defines the fact that Marshallian industrial districts (MIDs) as a model of production are more innovative than the whole national economy. In this paper we have measured the iMID effect with regard to other types of LPSs between 1991 and 2014 using a detailed database of 143,000 patents and utility models for the Spanish economy.

The paper introduces significant novelties compared to previous works based on the iMID effect (e.g. Boix and Galletto 2009; Boix, Galletto, and Sforzi 2019). The most important is that the territorial unit of analysis is not static at a point in time, but is dynamic, as it is subject to constant redefinition over time, both morphologically (changes in the geographical boundaries) and typologically (changes in the industrial characteristics).

For the first time, a complex causal analysis that goes beyond knowledge production functions has been introduced into the study of the iMID effect. For this, directed acyclic graphs have been used. This has made it possible to test the problems and validity of previous methods of analysis, clarifying what effective controls were being introduced, and controlling for new factors such as policy or innovation inertia. Although the great novelty, in this case, is the ability to simultaneously analyse the effects of the past and current types of LPSs and the transformations between types. For the estimation of the causal effects, ML methods have been used, one of which, the random forest, has proven to be especially flexible to be able to introduce these effects without being affected by the limitations of the traditional methods.

The results show that there have been very intense morphological and typological transformations in the MIDs and other types of LPSs in the period under analysis (1991–2014). Therefore, what is measured is the innovative intensity of the LPSs conditional on their being MID (or other types of LPSs), and not that of LPSs that do not change over time.

Using this approach, it is shown that the MIDs have maintained an innovative intensity (iMID effect) higher than the average of the LPSs of the Spanish economy throughout the period 1991–2014, although it seems to have decreased between 2008 and 2014 as a consequence of the 2008 crisis. The average causal effect is around 39% and is in the line with previous studies using static territorial units (Boix, Galletto, and Sforzi 2019). From this point of view and for this specific problem and period, it seems that the aggregated bias of using static territorial units is not excessive. However, when controlling for the inertia in the generation of innovations, the iMID effect for a current period t is reduced to 19.5% with the MIDs showing very high inertia.

The use of the type of the LPSs in current and lagged periods in the estimates has also made it possible to measure, for the first time, the effect of being MID (or other type of LPS) in the past on current innovative intensity, as well as those of the transformations between types of LPS. The results show that the effect of being MID in the previous period is, on average, as powerful as being MID in the current period, with the aggregate effect of being MID in both periods ranging around 27%–30% higher than the national average. The MIDs have proven to be, during the period under analysis, living innovation machines, whose competitive advantage is based on their ability to generate innovation.

Other typological transformations have also been shown to explain higher or lower innovation differentials. Throughout this process, a relevant result is that the transformation of a type of LPS into another type of LPS does not automatically imply a change in innovative capacity, and the type of origin continues to be highly relevant in explaining current innovation intensity, in other words, the ability of local communities to innovate change more slowly than their typology.

Although the results are robust and complement those of previous research on the iMID effect, this work has some limitations. The first is a technical limitation. It concerns the reproduction of the measurements beyond 2014, which requires updating the database and the 2021 LPSs in Spain, unworkable until the Census is fully processed. The second is a conceptual limitation. It concerns the causes of the typological transformation of local communities that have not been thoroughly investigated in this work. This limitation depends on the focus of this paper, which is methodological, while the search for the causes of the typological change is a matter of theoretical thinking, and it requires a work on its own which presupposes the availability of empirical research findings, such as those achieved in this paper. Future research will focus on this topic.

Notes

1. Other variables such as serendipity or the extreme effect of some business units would directly affect the innovative intensity without being visibly related to other covariates, which would not affect the causal estimation of the effects of the typology of LPS.
2. The sources of data are SABI and other business directories, Spanish National Institute of Statistics, Ministry of Education and Universities (Boix, Galletto, and Sforzi 2019).
3. For the year 2011, Boix et al. (2018) identification is used, which subdivides very large LLSs when necessary,adding 11 LPS to those identified by Sforzi and Boix (2019).

Acknowledgements

The authors would like to thank the two anonymous referees for helpful comments made on previous versions of this paper. The usual disclaimers apply.

Disclosure statement

No potential conflict of interest was reported by the author(s).

Funding

This work was supported by Ministerio de Ciencia, Innovación y Universidades (Grant Number RTI2018-095739-B-100 and PID2021-128878NB-100).

ORCID

Rafael Boix-Domenech http://orcid.org/0000-0003-0971-3464
Vittorio Galletto http://orcid.org/0000-0003-1709-8000
Fabio Sforzi http://orcid.org/0000-0002-3312-0995
Francesco Capone http://orcid.org/0000-0003-2000-3033

References

Apley, D., and J. Zhu. 2020. "Visualizing the Effects of Predictor Variables in Black box Supervised Learning Models." *Journal of the Royal Statistical Society. Series B: Statistical Methodology* 82 (4): 1059–1086. doi:10.1111/rssb.12377

Athey, S., and G. Imbens. 2019. "Machine Learning Methods That Economists Should Know About." *Annual Review of Economics* 11: 685–725. doi:10.1146/annurev-economics-080217-053433

Becattini, G. 1990. "The Industrial District as a Socio-Economic Notion." In *Industrial Districts and Inter-Firm Cooperation in Italy*, edited by F. Pyke, G. Becattini, and W. Sengenberger, 37–51. Geneva: International Institute for Labour Studies.

Becattini, G. 2004a. *Industrial Districts: A new Approach to Industrial Change*. Cheltenham: Edward Elgar.

Becattini, G. 2004b. *The Caterpillar and the Butterfly*. Firenze: Le Monnier.

Boix, R., F. Capone, and V. Galletto. 2022. "Searching for "Rare Diamonds"? Industrial Districts and Innovation in Spain and Italy." *Competitiveness Review: An International Business Journal* (Ahead-of-Print). doi:10.1108/CR-11-2021-0155.

Boix, R., and V. Galletto. 2009. "Innovation and Industrial Districts: A First Approach to the Measurement and Determinants of the I-District Effect." *Regional Studies* 43 (9): 1117–1133. doi:10.1080/00343400801932342

Boix, R., Sforzi, F., Galletto, V., and J. Llobet. 2018. *Los distritos industriales en España: 1991, 2001, y 2011*. IERMB and Universitat de València.

Boix, R., V. Galletto, and F. Sforzi. 2019. "Place-based Innovation in Industrial Districts: The Long-Term Evolution of the iMID Effect in Spain (1991–2014)." *European Planning Studies* 27 (10): 1940–1958. doi:10.1080/09654313.2019.1588861

Boix, R., and J. Trullén. 2010. "Industrial Districts, Innovation and I-District Effect: Territory or Industrial Specialization?" *European Planning Studies* 18 (10): 1707–1729. doi:10.1080/09654313.2010.504351

Breiman, L. 2001. "Random Forests." *Machine Learning* 45 (1): 5:32. doi:10.1023/A:1010933404324

Cooke, P. 2013. "Introduction: Complex Systems Integration, 'Emergence', and Policy Modularisation." In *Re-framing Regional Development: Evolution, Innovation and Transition*, edited by Cooke Ph, 1–26. London: Routledge.

Dei Ottati, G. 2018. "Marshallian Industrial Districts in Italy: The end of a Model or Adaptation to the Global Economy?" *Cambridge Journal of Economics* 42 (2): 259–284. doi:10.1093/cje/bex066

Galletto, V., and R. Boix-Domènech. 2014. "Industrial Districts, Technological Innovation and I-District Effect: A Question of Volume or Value?" *Investigaciones Regionales - Journal of Regional Research* 30: 27–51.

Griliches, Z. 1992. "The Search for R&D Spillovers." *The Scandinavian Journal of Economics* 94: 29–47. doi:10.2307/3440244

Hägerstrand, T. 1970. "What About People in Regional Science?" *Papers of the Regional Science Association* 24 (1): 6–21. doi:10.1007/BF01936872

Hothorn, T., K. Hornik, and A. Zeileis. 2006. "Unbiased Recursive Partitioning: A Conditional Inference Framework." *Journal of Computational and Graphical Statistics* 15 (3): 651–674. doi:10.1198/106186006X133933

Lazzeretti, L., and F. Capone. 2017. "The Transformation of the Prato Industrial District: An Organisational Ecology Analysis of the co-Evolution of Italian and Chinese Firms." *The Annals of Regional Science* 58 (1): 135–158. doi:10.1007/s00168-016-0790-5

Loasby, B. 2002. "Expectations, Uncertainty and Beliefs in the Marshallian Analysis of Markets." *Revue D'Économie Politique* 112 (5): 659–670. doi:10.3917/redp.125.0659

Marshall, A. 1920. *Principles of Economics*. 8th ed. London: MacMillan.

Molnar, C. 2022. *Interpretable Machine Learning. A Guide for Making Black Box Models Explainable*. Leanpub: Victoria.

Openshaw, S., M. Charlton, and A. Craft. 1988. "Searching for Leukaemia Clusters Using a Geographical Analysis Machine." *Papers of the Regional Science Association* 64: 95–106. doi:10.1111/j.1435-5597.1988.tb01117.x

Openshaw, S., and P. Taylor. 1981. "The Modifiable Areal Unit Problem." In *Quantitative Geography: A British View*, edited by N. Wrigley and R. Bennett, 60–69. London: Routledge.

Pearl, J., and D. Mackenzie. 2018. *The Book of why*. London: Penguin Books.

Sforzi, F. 2012. "From Administrative Spatial Units to Local Labour Market Areas." In *Defining the Spatial Scale in Modern Regional Analysis. New Challenges from Data at the Local Level*, edited by E. Fernández-Vázquez and F. Rubiera Morollón, 3–21. Berlin-Heidelberg: Springer.

Sforzi, F., and R. Boix. 2015. "What About Industrial District(s) in Regional Science." *"Investigaciones Regionales - Journal of Regional Research* 32: 61–73.

Sforzi, F., and R. Boix. 2019. "Territorial Servitization in Marshallian Industrial Districts: The Industrial District as a Place-Based Form of Servitization." *Regional Studies* 53 (3): 398–409. doi:10.1080/00343404.2018.1524134

Swann, P. 2009. *The Economics of Innovation: An Introduction.* Cheltenham: Edward Elgar.

Trullén, J., R. Boix, and V. Galletto. 2013. "An Insight on the Unit of Analysis in Urban Research." In *Handbook of Research Methods and Applications in Urban Economies*, edited by P. Kresl and N. Sobrino, 235–264. Northampton: Edward Elgar.

Varian, H. 2014. "Big Data: New Tricks for Econometrics." *The Journal of Economic Perspectives* 28 (2): 3–28. doi:10.1257/jep.28.2.3

Appendix

Annex 1.

	AGR_t	Innovative intensity	MID_t	MID_{t-1}	Stock of R&D public	OSE_t	OSE_{t-1}	Stock of R&D private	MLF_t	MLF_{t-1}	CON_{t-1}	CON_{t-1}	BSE_t	BSE_{t-1}
AGR_t	0.71	-0.26	-0.57	-0.45	-0.02	-0.31	-0.24	-0.05	-0.25	-0.14	-0.22	-0.15	-0.05	-0.03
AGR_{t-1}		-0.24	-0.42	-0.6	-0.03	-0.24	-0.39	-0.06	-0.16	-0.16	-0.11	-0.22	-0.05	-0.05
Innovative intensity			0.24	0.26	-0.01	-0.07	-0.06	0.08	0.16	0.14	-0.02	-0.02	0.02	0
MID_t				0.68	-0.06	-0.25	-0.17	0.01	-0.2	-0.09	-0.18	-0.11	-0.04	-0.03
MID_{t-1}					-0.04	-0.18	-0.28	0.03	-0.02	-0.11	-0.12	-0.16	0.01	-0.03
Stock of R&D public						0.09	0.07	0	-0.02	-0.01	0.03	0.03	0.05	0.03
OSE_t							0.59	-0.04	-0.11	-0.06	-0.1	-0.01	-0.02	0
OSE_{t-1}								0.01	0.03	-0.07	-0.03	-0.1	0.01	-0.02
Stock of R&D private									0.14	0.05	-0.04	0.01	0.06	0.03
MLF_t										0.52	-0.08	-0.03	-0.02	0.01
MLF_{t-1}											-0.05	-0.04	-0.01	-0.01
CON_t												0.55	-0.02	0.01
CON_{t-1}													-0.01	-0.01
BSE_t														0.45

Colour scale: 1, 0.8, 0.6, 0.4, 0.2, 0, -0.2, -0.4, -0.6, -0.8, -1

Figure A1. Correlation matrix. Type of LPS: AGR: Agricultural and mining; CON: Construction; MID: Marshallian industrial districts; MLF: Manufacturing LPSs of large firms; BSE: Business services LPSs; OSE: Other services LPSs. Transformations include endogenous evolution or merging/division in another LPS with another dominant specialization. The suffixes *t* and *t-1* express the current and the previous period, respectively.

Annex 2. *Marginal effects by type of LPS* from random forest predictions and ALE values.*

	Model 1			Model 2			Model 3			Model 4		
Type of LPS	Mg effect[(a)]	% mg effect[(b)]	z[(c)]	Mg effect[(a)]	% mg effect[(b)]	z[(c)]	Mg effect[(a)]	% mg effect[(b)]	z[(c)]	Mg effect[(a)]	% mg effect[(b)]	z[(c)]
AGR_t	-112.8870	−0.4569	***	−116.8171	−0.4728	***	−107.1295	−0.4337	***	−87.0812	−0.3524	**
CON_t	−10.8289	−0.0438		−10.7660	−0.0436	*	3.4535	0.0140		−2.45593	−0.0099	
MID_t	97.1371	0.3931	***	48.0767	0.1946	***	31.1798	0.1262	***	32.6003	0.1319	*
MLF_t	98.8352	0.4000	***	74.3157	0.3008	***	82.1903	0.3327	***	87.3071	0.3534	*
BSE_t	7.9741	0.0323	**	28.9336	0.1171	*	34.6549	0.1403	**	19.9785	0.0809	*
OSE_t	−39.3912	−0.1594	**	−37.5450	−0.1520	***	−26.8889	−0.1089	***	−24.7450	−0.1001	*
AGR_{t-1}							−9.9836	−0.0404	**	−8.2102	−0.0332	
CON_{t-1}							−12.7494	−0.0516		−7.7090	−0.0312	
MID_{t-1}							35.4480	0.1435	***	32.2584	0.1306	
MLF_{t-1}							69.3400	0.2807	***	83.0320	0.3360	
BSE_{t-1}							−8.2235	−0.0333		−4.6674	−0.0189	
OSE_{t-1}							−18.9817	−0.0768	**	−16.5966	−0.0672	
Controls	R&D private, R&D public			LPS_{t-1}			LPS_{t-2}			LPS_{t-2}, LPS $Size_t$, LPS $Size_{t-1}$		
RMSE		390.81			395.89			366.52			368.29	
R2		0.1439			0.1216			0.1360			0.1276	
R2-cor[(d)]		0.3820			0.3487			0.3697			0.3661	
Observations		3046			3046			2342			2342	
Mean y		247.08			247.08			247.01			247.01	

*Type of LPS: AGR = Agricultural and mining; CON = Construction; MID = Marshallian industrial districts; MLF = Manufacturing LPSs of large firms; BSE = Business services LPSs; OSE = Other services LPSs. Transformations include endogenous evolution or merging/division in another LPS with another dominant specialization. The suffixes *t* and *t*-1 express the current and the previous period, respectively.

[(a)]% Marginal effect expressed as number of innovations per million of persons employed.

[(b)]% mg effect is the average marginal conditioned effect divided by the mean of the predicted innovation per million employees.

[(c)]z = percentage of increase in the mean squared error (average decrease of accuracy in the predictions if the feature is excluded from the estimates). The higher the value the better: *** z-value above 30%, ** z-value above 20%, * z value above the 10%. Negative values of z are interpreted as a very low predictive ability (a random feature performs better than the excluded feature).

[(d)]Pearson correlation between the real and predicted outcomes.

The mechanisms underlying the emergence of innovation ecosystems: the case of the AI ecosystem in Montreal

Nasrin Sultana, Ekaterina Turkina and Patrick Cohendet

ABSTRACT
Scholars have increasingly been using ecosystem as a concept within and beyond social science, but less is known about how ecosystems emerge. In this study, we investigate the context of the Artificial Intelligence (AI) ecosystem in Montreal to understand the mechanisms underlying the emergence of innovation ecosystems. Building on the work of Ostrom and the literature on innovation commons and conducting content analysis and network analyses, we find empirical evidence for a bottom-up approach in the emergence of the AI ecosystem in Montreal. We find that the main mechanism underlying the emergence of innovation ecosystems in Montreal is the articulation of a series of innovation commons by commoners. Our findings have implications for understanding the importance of emerging technologies and the digitalization of industries and for identifying regional innovation capabilities.

1. Introduction

Research on innovation ecosystems received increasing attention from scholars and practitioners in the last decade. An innovation ecosystem provides an environment for the actors with a wealth of technical expertise, business experience, and access to capital that supports innovation (Adner 2006; Autio and Thomas 2014; Granstrand and Holgersson 2020; Jackson 2011). Scholars have examined the origin of innovation ecosystems, trying to identify the mechanisms underlying the emergence of these innovation ecosystems. These mechanisms reflect the efforts and actions of different actors in their intention to innovate (Lacouture et al. 2015; Tsujimoto et al. 2018).

However, there is not yet a conceptual consensus on the theoretical and empirical understanding of how an ecosystem emerges. The literature on the emergence of ecosystems is fragmented with one stream arguing in favour of the bottom-up approach insisting on the efforts and actions of diverse actors to achieve collective goals (Colombo et al. 2019; Gifford, McKelvey, and Saemundsson 2021; Lenkenhoff et al. 2018), another stream arguing in favour of the top-down approach by highlighting diverse processes and programmes, such as financing projects and facilitating spin-offs, put in place by a given authority to create the ecosystem (Romero and Molina 2011; Thomas, Faccin, and Asheim 2021), and yet another stream arguing in favour of a combination of both approaches (Sun et al. 2019; Fu 2021). The fragmentation in the research on innovation ecosystem arises primarily because of ambiguous terminologies used interchangeably across different types of ecosystems without sufficient theoretical support (Smorodinskaya et al. 2017). Accordingly, there has been a consistent call to study ecosystems, align with theories, and ascertain how ecosystems emerge (Ritala and Gustafsson 2018). To answer this call, we investigate the context of the AI ecosystem in Montreal[1] to understand the mechanisms underlying the emergence of innovation ecosystems.

In less than two decades, Montreal has become one of the most active ecosystems of innovation in AI in the world. In terms of scientific research, the AI academic community of Montreal is now the largest in North America. In particular, the city hosts the Montreal Institute for Learning Algorithms (MILA) – a research institution internationally recognized for its significant contributions to machine learning. Another key institution is IVADO – an advanced multidisciplinary centre for professionals and researchers to develop resources and expertise in data science, operational research, and AI. Montreal is also at the forefront of responsible AI with Montreal Declaration for responsible AI development. The development of AI in Montreal has found in particular a rich milieu of applications in the domain of health. One of the main initiatives is the creation of the CHUM School of Artificial Intelligence in Healthcare (SAIH) in 2018 at the Montreal University Hospital Center (CHUM), in conjunction with the University of Montreal. The objective of the SAIH is to support and assist healthcare system stakeholders in the adoption, implementation, and promotion of AI. The CHUM also initiated the CITADEL project to facilitate the access of data and use of AI. The concentration of talents and expertise in the domain of health and AI has led to collaborations between different actors in hospitals, universities, and research centres and to the development of numerous Montreal-based companies using AI in healthcare. One of the most active companies is Imagia, which has invested in deep learning to analyze medical imaging data to predict cancer patient outcomes with increasing accuracy.

With regards to the issue of factors and antecedents to the formation of the innovation ecosystem in AI in Montreal, it appears that none of the traditional top-down modes of explanation in terms of regional structural preconditions of the formation of an ecosystem of innovation applies to the case of AI in Montreal. There was no pre-existing anchor firm or dominant firm in the domain, such as in the case of the videogame ecosystem of innovation in Montreal, which has been driven by the anchor firm Ubisoft (Cohendet 2022). There was a concentration of firms and labs in information and communication technology (ICT) in the city leading to a strong ICT cluster. However, the cluster was not comparable, by size or strength, to other main ICT clusters in Canada such as Toronto, Vancouver, and Calgary. Additionally, the activities were not focused on innovation in AI. There was no active and deliberate initial public funding to build an ecosystem of innovation, such as in the case of the massive funding in biotech academic centres in California (Feldman, 2005) that led to the formation of a myriad of university spin-offs as knowledge externalities, or such as the development of the aeronautic ecosystem of innovation in Toulouse (Longhi, 2005) heavily subsidized by the French government.

The federal government of Canada and the provincial government of Québec have recently been increasingly investing in AI. Among the flagship institutions supported by public funds is IVADO – an advanced multidisciplinary centre for professionals and researchers to develop resources and expertise in data science, operational research, and AI. To ensure the responsible use of AI, the provincial government of Québec supported the creation of the International Observatory on the Societal Impacts of AI and Digital Technology (OBVIA). However, such financial support was not at the origin of the formation of AI ecosystem but came later in the process to reinforce already existing successful initiatives.

Therefore, we aim at understanding the mechanisms underlying the emergence of such a vibrant ecosystem of innovation in AI in Montreal. Considering the observations mentioned above, our view is that the emergence of the AI innovation ecosystem in Montreal is issued from a more systemic and multi-actor perspective that incorporates non-firm actors, non-local sources, future expectations and visions, and inter-path relations (Hassink, Isaksen, and Trippl 2019; Jolly, Grillitsch, and Hansen 2020). There are key local individual actors, who were not only at the origin of the emergence of the ecosystem of innovation in AI in Montréal but also through their collective actions, orchestrated the different steps of development of the ecosystem and supported the institutionalization process of new path creation that helped to push the industry forward (Doloreux and Turkina 2021). These key actors, called commoners, facilitate the articulation of a series of 'innovation commons' that enable knowledge to be progressively revealed, enhanced, nurtured, interpreted, and enacted collectively, leading to the emergence of innovation ecosystems (Allen and Potts 2016). Therefore, the main mechanism underlying the emergence of innovation ecosystems in Montreal is the articulation of a series of innovation commons by commoners. Based on the actions of commoners in different steps of the development of the ecosystem, we identified three sub-mechanisms: orchestration of social commons, orchestration of symbolic commons, and orchestration of knowledge commons.

We aim at contributing to the literature on ecosystems by using the notion of innovation commons to highlight that the bottom-up initiatives of commoners are at the origin of the formation of an innovation ecosystem. We also aim to improve the understanding of the emergence of innovation ecosystems, and thereby, bridge the gap between different streams of research on innovation ecosystems. The findings also provide means to support the development of innovation ecosystems. The paper is structured as follows: we dedicate the next section to the theoretical background. We present the research design in section three and the findings in section four. In sections five and six, we discuss the implications of our findings and conclude by providing future research directions.

2. Theoretical background and hypotheses

Although scholars are becoming increasingly interested in studying innovation ecosystems, there remain conceptual ambiguity, methodological challenges, and a lack of rigorous assessment guidelines (Oh et al. 2016; Ritala and Gustafsson 2018). Following the assemblage theory by Manuel DeLanda (2016, 2019), it can be said that ecosystem is a concept with knobs that can be set to define different ecosystems. The knobs can be value propositions, structures, scopes, or functions. For instance, actors in an innovation ecosystem are defined according to scope, affiliation, or organizational boundaries, and the interaction level of an ecosystem is defined by the operational environment such as industry alignments or country economic dynamics (Ngongoni, Grobbelaar, and Schutte 2022, 5). Accordingly, we can say that in an innovation ecosystem, the collective goal or the knob is set around the activities related to innovation. In this section, we discuss the literature on innovation commons and the role of commoners in innovation ecosystems and build hypotheses on the mechanisms in different steps of the emergence of innovation ecosystems.

2.1. Innovation Ecosystem: Innovation commons and role of commoners

The concept of innovation ecosystems has gained considerable research interest from scholars in different domains during recent years (de Vasconcelos Gomes et al. 2018; Faissal Bassis and Armellini 2018; Granstrand and Holgersson 2020). The concept has its root in the related concept of business ecosystems as used by Moore (1993). Moore defined business ecosystem as a community of interacting organizations and individuals including customers, lead producers, competitors, and other stakeholders (Moore 1993). The concept of innovation ecosystems started to gain currency after Adner (2006) defined an innovation ecosystem as the collaborative arrangements through which firms combine their individual offerings into a coherent, customer-facing solution. In an innovation ecosystem, the focus is usually on the introduction of a new product or service or a new way to create value for customers by introducing a new or changing an existing business model (Yaghmaie and Vanhaverbeke 2019). The functional goal of actors or entities in innovation ecosystems is to enable technology development and innovation (Jackson 2011). The notion of the ecosystem of innovation has been increasingly discussed in the management literature (Iansiti and Levien 2004; Adner 2006; de Vasconcelos Gomes et al. 2018), which argues that the dynamics of the development of innovation in a given domain result from heterogeneous interdependencies between different stakeholders including informal groups and communities.

In this study, we define the groups of key passionate individuals as commoners who are linked together by a common goal, a common cognitive framework, and a shared understanding of their work. Commoners, as members of a community (Ostrom 1990, 2009) or a core group of the professional community (Amin and Roberts 2008), have access as well as the right to use, preserve, and develop the common belonging to the community. Ostrom defines commons as a 'collective action governance mechanism over a common pool resource shared by the members of a community ("the commoners") who jointly manage the use and access to this resource as well as its preservation or development' (Zimmermann 2020). While Ostrom refers to a resource environment, in particular natural resources, by commons, scholars have extended the scope of the term to refer to other resources including but not limited to knowledge, culture, and innovation (Allen and Potts 2016).

Allen and Potts highlighted, through the notion of innovation commons, that the institutional origin of new technologies may be related to 'self-organizing groups of technology enthusiasts who develop effective governance rules to pool distributed information resources. The "innovation commons" alleviate uncertainty around a nascent technology by pooling distributed information about uses, costs, problems, and opportunities' (Allen and Potts 2016, 1), which reveals innovative opportunities and reduces uncertainty in innovative processes. Our view is that commoners facilitate the articulation of a series of 'commons', which enables knowledge to be progressively revealed, enhanced, nurtured, interpreted, and enacted collectively (Allen and Potts 2016; Cohendet 2022; Helfrich and Haas 2009). We argue that the mechanisms underlying the emergence of innovation ecosystems are the bottom-up initiatives of commoners leading to the articulation of a series of commons, namely, social commons, symbolic commons, and knowledge commons.

2.2. Mechanisms underlying the emergence of innovation ecosystems

A mechanism results in the interaction between human agents, intervention, and structures and evolves within an open space-time and social system of relationships (Lacouture et al. 2015). According to the literature, the mechanisms underlying the emergence of innovation ecosystems result from the efforts and actions of actors (Lacouture et al. 2015; Tsujimoto et al. 2018). The theories that explain the formation of high-technology innovation mostly rely on the analysis of top-down initiatives such as the public funding of research projects (Sun et al. 2019), the key role of an anchor firm (Iansiti and Levien 2004), the positive externalities of research resulting from a dense innovation system (Romero and Molina 2011), or the spin-offs from university research (Thomas, Faccin, and Asheim 2021). In such a perspective, these top-down initiatives mostly rely on the organization of specific local clusters of activities to trigger the innovation processes they initiate. These theories do not generally attribute the rise of high levels of entrepreneurship and innovativeness to cooperation and collaboration, especially a collaboration mode that is issued from bottom-up initiatives carried out by an informal community.

Innovation ecosystems can be formed autonomously when a sufficiently large group of actors aligns strategic interests (Tsujimoto et al. 2018). For example, scholars have studied the role of diverse informal communities in the dynamics of knowledge formation at the local level (Bathelt, Malmberg, and Maskell 2004; Cole and Barberá-Tomás 2014; Von Krogh and Geilinger 2014). Similarly, scholars have highlighted the role of grassroots initiatives in tackling some of the challenges that drive transformational change (Grandadam, Cohendet, and Suire 2022; Smith and Stirling 2018). However, this rich literature does not specify the precise initiatives of local communities of passionate individuals to orchestrate the process of innovation or the mechanisms underlying the emergence of innovation ecosystems.

Understanding the mechanisms of innovation ecosystems will enable us to better perceive not only the activities and relationships between actors but also changes in the structure of the ecosystem (Bettiol and Sedita 2011; Simmonds et al. 2021). Below, we build hypotheses to highlight the bottom-up approach and the relevant mechanisms underlying the emergence of innovation ecosystems.

2.2.1. Orchestration of social commons

Commoners play important roles in the emergence and development of radical ideas leading to the formation of rich and vibrant innovation ecosystems in a location that does not necessarily have a prior history or antecedent in the field or sector (Cohendet 2022). Initially, based on the openness and active cross-fertilization of a network of like-minded professionals, commoners contribute to the formation and maintenance of social commons (Amin and Roberts 2008). The commoners are strongly involved as a core group of active members in different types of communities. They share the ideas, skills, and various types of knowledge necessary to achieve their objectives

(Kirwan, Dawney, and Brigstocke 2015). The efforts undertaken by members of the diverse communities are eventually translated into social commons (Helfrich and Haas 2009). The main attributes inherent to social commons are consensus, equity, moral legitimacy, and transparency in decision-making (Willis 2012).

The efforts undertaken by the professional community in Montreal, can therefore, be translated as the development of a *social commons* focused on the active opening of personal networks and the conceptualization and promotion of relationships between the diverse members of the community. Following Hassink, Isaksen, and Trippl (2019) and Jolly, Grillitsch, and Hansen (2020), our view is that the emergence of the AI innovation ecosystem in Montréal is issued from a more systemic and multi-actor perspective that incorporates non-firm actors, non-local sources, future expectations, visions, and inter-path relations. There are key local passionate actors who connect actors from diverse backgrounds, leading to the orchestration of *social commons* and contributing to the emergence of the ecosystem of innovation in AI in Montreal. Therefore, we argue:

> H1: Commoners are the facilitators of linkages between actors, leading to the orchestration of social commons and contributing to the emergence of innovation ecosystems.

2.2.2. Orchestration of symbolic commons

In an ecosystem, a collective of firms or organizations is connected by a shared interest to materialize their product or service innovation (Lenkenhoff et al. 2018). For example, innovation ecosystems are defined as complex relationships formed by an interdependent network of actors with a functional goal to enable value creation and appropriation through innovation and technology development (Ashton 2009). Innovation ecosystems are not controlled by one large focal firm, but activities in an ecosystem centre around the firm, activity, or technology that influences an ecosystem's responses to externalities and its evolution (Ngongoni, Grobbelaar, and Schutte 2022). There are different activities and functions in an ecosystem involving resources, institutions, contexts, and technologies that connect relevant actors. The structure of an ecosystem allows participants to collaborate and solve coordination challenges without following a full hierarchy (Jacobides, Cennamo, and Gawer 2018).

With the emergence of an innovation paradigm and the opening of innovation processes to society, companies, technical schools, and research institutes are no longer the only relevant agents in the process of innovation (Domanski, Howaldt, and Kaletka 2020). Innovations occur within a complex context of interaction between different actors (Faissal Bassis and Armellini 2018). The diversity of the relevant actors and the structure of their network affect innovation in a region (Innocenti et al. 2022). The collaborative environment required in innovation ecosystems often results from bottom-up initiatives carried out by an informal group of passionate individuals. These individuals frame the ideas from collective endeavour in a shared vision and develop symbolic commons to facilitate the process of innovation (Bettiol and Sedita 2011; Bowers 2004). The decentralized structure of the community of actors leads to the development of a shared understanding and orchestration of symbolic commons (Bowers 2004; Hassink, Isaksen, and Trippl 2019). Therefore, we argue:

> H2: A decentralized structure of the community of actors leads to the orchestration of symbolic commons and contributes to the emergence of innovation ecosystems.

2.2.3. Orchestration of knowledge commons

Ecosystems are increasingly being studied as a collaborative context where actors with complementary resources come together to create and capture value (Adner and Kapoor 2010; Radziwon, Bogers, and Bilberg 2017). For example, ecosystems in a region can be conceptualized as complex systems in which individuals and organizations have multiple connections between them as well as the ability to learn and adapt to external and internal changes (Ashton 2009). There are complex and dynamic interactions in an ecosystem involving individuals, business organizations, government agencies, industrial players, universities, and research institutes (Lenkenhoff et al. 2018). It is not necessary for the background of actors to be constrained by the scope of the ecosystem and vice versa; it is the collective goal that defines the scope or nature of an ecosystem (Adner and Kapoor 2010). Creativity, connectivity, collaboration, and community are crucial competencies of the actors involved in an ecosystem (Karakas 2009). The collaborative activities between actors play critical roles in all phases of successful innovation (Arena et al. 2017; Radziwon, Bogers, and Bilberg 2017).

Innovation ecosystems are seen as the crossing boundaries of a single industry or sector (Autio and Thomas 2014) that result in innovation (Valkokari 2015; Paasi et al. 2022). The boundaries of an ecosystem are related to the nature of the value proposition as well as to the structure of interdependencies (Adner and Kapoor 2010). The network of relationships between actors in innovation ecosystems allows them to exchange knowledge (Amin and Roberts 2008; Kirwan, Dawney, and Brigstocke 2015). For example, the actors coming from different domains establish bridges between different communities and sustain knowledge creation (Bettiol and Sedita 2011). Accordingly, actors in innovation ecosystems come from diverse backgrounds and bring complementary knowledge and resources across boundaries, which leads to the orchestration of knowledge commons. Therefore, we argue:

> H3: The collaboration between actors from diverse backgrounds and across boundaries leads to the orchestration of knowledge commons and contributes to the emergence of innovation ecosystems.

To sum up, in emerging innovation ecosystems, the bottom-up initiatives of commoners lead to the articulation of a series of innovation commons. At first, commoners orchestrate social commons by connecting actors from diverse backgrounds. Next, the decentralized structure of the community of actors leads to the orchestration of symbolic commons. Finally, the orchestration of knowledge commons occurs due to the collaboration between actors across boundaries and their attendant knowledge. Figure 1 shows the relationship between commoners, organizations, ecosystems, and communities in a region.

In the following sections, we elaborate on the research setting, conduct the analysis, report the findings, and discuss the implications of our findings.

3. Research design

3.1. Research setting: the AI ecosystem in Montreal[2]

The AI ecosystem in Montreal provides a solid ground to investigate how innovation ecosystems emerge. An innovation ecosystem can generate new output without using input from the originator, is resilient, and attract new external actors to enhance the innovative potential of the system (Saxenian 2018). These attributes are present in the AI ecosystem in

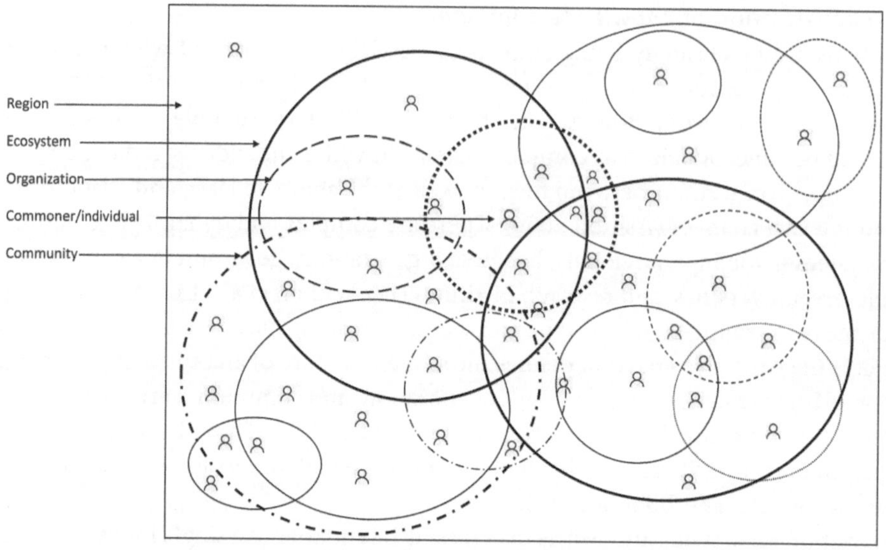

Figure 1. The relationship between commoners, organizations, communities, and ecosystems in a region.

Montreal. With regards to the issue of factors and antecedents to the formation of the innovation ecosystem in AI in Montréal, none of the traditional modes of explanation in terms of regional structural preconditions apply. For instance, there was no pre-existing anchor or dominant firm in the domain. Besides, the region has become home to one of the world's largest deep-learning university communities, and global leaders including Google, Microsoft, and Facebook are setting up research labs in the region.[3]

A representation of the AI ecosystem, using the data available from Montreal International,[4] is shown in Figure 2. The current AI ecosystem of Montreal consists of more than 400 different types of organizations including business firms, research institutions, and governmental units. As a result of this exceptionally high concentration of scientific expertise in AI, major industrial players from other countries have been recently investing in Montréal's AI industry.

The dynamics of the innovation ecosystem in AI in Montreal showed a broad range of partnerships with famous global companies including Google, Facebook, Microsoft, and Samsung. Nonetheless, many of the AI companies in Montreal are small units; over 90% of Montreal's firms in AI are startups, and a large part of the vibrancy of AI activities in the city comes from informal communities of passionate engineers, techno-geeks, or individual entrepreneurs. The concentration of talents and expertise in the domain of AI and health in Montreal has led to collaborations between different actors and the development of numerous Montreal-based companies using AI in healthcare. Given the context, the majority of the actors we studied have collaborative relationships in AI and healthcare.

3.2. Methods and data

In this study, we adopt an abductive approach – combining qualitative content analysis and social network analysis to identify the connection between commoners and organizations

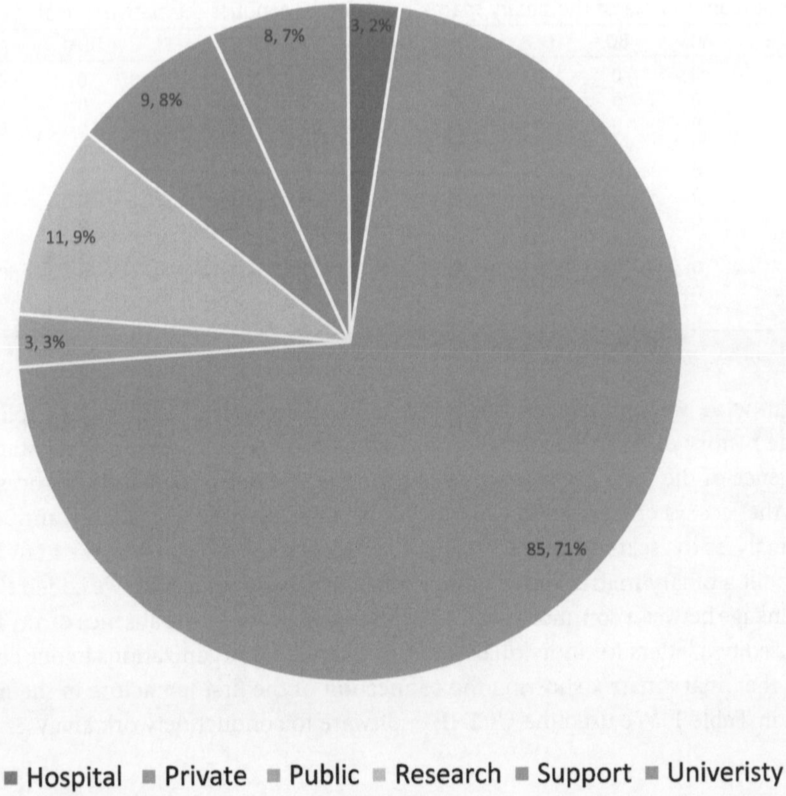

Figure 2. Distribution of actors by type in the AI ecosystem in Montreal.

and unfold the story of ecosystem emergence from bottom-up approaches (Boréus and Bergström 2017; Borgatti, Everett, and Johnson 2018; Sætre and Van de Ven 2021). We conducted semi-structured interviews with five experts in the AI ecosystem in Montreal to validate our findings. The interviews were typically from 45 to 55 min long. The abductive approach is fitting in this research because generating explanations of the emergence of innovation ecosystems by exploring the AI ecosystem in Montreal involved iteration between the analyses and findings: first, we observe the initiatives of commoners in the context of the emergence of the AI ecosystem in Montreal; next, both the findings from network analysis and the conversation with experts confirm our observation (Sætre and Van de Ven 2021).

The data used in this research come primarily from secondary sources. We compiled data from the information and documents available online for the period from August 2021 to July 2022. First, we prepared a list of commoners and organizations in the AI ecosystem in Montreal based on the information available from Montreal international and complemented the list using information from other sources. All data used to describe and analyze the AI ecosystem in Montreal come from the sources mentioned in Appendix A. Our final sample includes 19 commoners and 102 organizations that have at least one linkage. Next, we consulted the websites of commoners and organizations; we identified the linkages by looking into whether a commoner is connected to more than one organization and whether an organization has partnership connections with other organizations. Finally, we interviewed the experts to validate our findings.

Table 1. A random sample of the binary matrix prepared to conduct the network analysis.

Name of actors	ADB	BO	DEP	EA	IMG	KM	LS	PL	UDM	YB	...
ADV	0	0	0	0	0	0	1	0	0	0	
BO	0	0	0	0	0	0	1	0	0	0	
DEP	0	0	0	0	0	1	0	0	0	0	
EA	0	0	0	0	0	1	1	0	0	0	
IMG	0	0	0	0	0	0	1	0	0	0	
KM	0	0	0	0	0	0	1	0	0	0	
LS	0	0	0	0	0	1	0	1	0	0	
PL	0	0	0	0	0	0	0	0	0	0	
UDM	0	0	0	0	0	1	1	1	0	0	
YB	0	0	0	0	0	1	1	0	0	0	
...											

To begin with, we conducted a qualitative content analysis of the data we collected to analyze the history of AI in Montreal. The analysis provides us with an understanding of the emergence of the AI ecosystem as well as the role of commoners in the process. Next, we map the ecosystem based on partnership linkages between actors and conduct network analyses to ascertain the structure of the ecosystem. To conduct the network analyses, we built a binary matrix showing the connection between actors: we coded the existence of a linkage between commoners and organizations as 1, and the absence of any linkages as 0. We used two letters for individuals and three letters for organizations in our coding. A sample of the binary matrix showing the connection of the first ten actors in the matrix is presented in Table 1. We used the UCINET software to conduct network analyses.

4. Analysis and findings

We track down the history of AI in Montreal since the 80s and identify the commoners crucial for the emergence of the AI ecosystem in Montreal by conducting qualitative content analysis. Below, we elaborate on our findings of different phases and mechanisms underlying the emergence of the AI ecosystem in Montreal.

4.1. Findings from the content analysis: the emergence of the AI ecosystem in Montreal and the articulation of commons

In Montreal, initially, the commoners emerged as a core group from a professional community and contributed to the orchestration of the social commons based on the openness and active cross-fertilization of a network of like-minded professionals in AI. Next, in the form of epistemic community, the actors elaborate on a shared understanding – the symbolic commons. Finally, as communities of practice, the actors orchestrated the knowledge commons. Table 2 provides a summary of the emergence of the AI ecosystem in Montreal in terms of the periods, communities involved, and commons in different phases.

4.1.1. Phase I: the development of a professional community in AI and the orchestration of social commons

During the 1980s, following a period of disenchantment with the possibilities of AI and expert systems in the 1970s, a group of passionate researchers working in these fields during the 1980s was convinced of the possibilities of breakthroughs in these domains.

Table 2. The emergence of the AI ecosystem in Montreal.

Phase	Period	Community	Commons
Phase-I	1980–2000	Professional community	Social commons
Phase-II	2001–2010	Epistemic community	Symbolic commons
Phase-III	2011–2022	Communities of practice	Knowledge commons

Under the influence of Geoffrey Hinton, who was working on artificial neural networks modelled after the human brain, now considered among the 'godfathers of deep learning',[5] a professional community of researchers progressively formed in Montreal. Among the members of the professional community, were Yann LeCun, a researcher in AI considered one of the inventors of deep learning, and Yoshua Bengio, a professor in the Department of Computer Science and Operations Research, Université de Montréal.

One of the key individuals in Montreal is Yoshua Bengio who founded the Montreal Institute for Learning Algorithms (MILA), in collaboration with the University of Montreal and McGill University, while serving as the scientific director of the Institute for Data Development (IVADO). In 2000, Bengio made a major contribution to natural language processing with the paper 'A Neural Probabilistic Language Model'. The approach has led to a major shift in machine translation and natural language understanding systems. Members of this community, who were fully convinced of the potential of deep learning and visual recognition, combined specialized knowledge acquired through extended periods of research and training and created institutional trust based on standards of professional conduct. By sharing common experiences and values, this professional community created social commons (Helfrich and Haas 2009).

4.1.2. Phase II: the development of an epistemic community in AI and the orchestration of symbolic commons

At the beginning of the 2000s, despite some significant advances in terms of publications in the domain, the group of commoners in Montréal was experiencing a dry period for AI. Over-promising marketing efforts generated expectations that the products could not meet. At that time, only a few groups of researchers were working on AI in Canada and internationally, and Canadian researchers were attracted by sky-high salaries in Silicon Valley. The brain drains to Silicon Valley attracted many, but a pool of die-hard AI researchers including Yoshua Bengio, Yann LeCun, Joelle Pineau, and some other key researchers in the field decided at the last minute to stay in Quebec. The perseverance and determination of the group of commoners laid the groundwork for the Montreal AI renaissance in the 2010s, attracting governmental, academic, and industrial support, that lead to the foundation of the Montreal Institute for Learning Algorithms (MILA), the Institute for Data Development (IVADO), Element AI, an AI intelligence incubator that aims at turning AI research into real-world applications for deep learning technology, etc.

This was a turning point that led the commoners from moving and sharing ideas and reinforcing professional relationships to focusing on their visionary projects and finding ways to test their innovative ideas. At that time, the main cognitive objective of the group shifted toward the elaboration of a symbolic common in the form of a vision accompanied by a declaration of intent, expressing the breaking of established rules, to collectively develop a world-class ecosystem of innovation in AI in Montréal. The group of researchers as commoners were at the origin of the formation of a new community in Montreal that can be described as epistemic, according to the definition of Cowan, David, and Foray (2000, 234). The community developed significant efforts in bringing together important actors who adhered deeply to the values and the vision of the commoners, thereby creating shared values or symbolic commons (Cohendet 2022).

4.1.3. Phase III: the development of communities of practice and the orchestration of knowledge commons

Finally, the community gained momentum as commoners from diverse communities of practice start collaborating to manage, preserve, and enrich the commons in Montreal. Building on the existing social commons and symbolic commons, the main actors in AI in Montréal progressively turned their focus to the application of scientific research in AI and innovations in industrial domains. By creating a pool of distributed knowledge and application of such knowledge leading to innovation, the communities of practice created knowledge commons. The network of relationships between actors allows them to exchange knowledge, and the actors coming from different domains establish bridges between different communities and sustain knowledge creation that would benefit the whole community.

Researchers and entrepreneurs in Montreal are emerging and working to shape the future of AI in Montreal. Hugo Larochelle leads the Google Brain practice in Montreal. Joelle Pineau teaches at McGill University and leads the Facebook Artificial Intelligence Research (FAIR) Lab at Facebook. Jean-François Gagné, co-founder and CEO of Element AI, brings the latest AI discoveries to companies. Alexandre Le Bouthillier and Nicolas Chapados in collaboration with Yoshua Bengio founded Imagia. Imagia contributes to the growth and prominence of Quebec's AI sector. Beginning in 2018, the government of Canada and Quebec recognized the importance of these developments and announced their support for the creation of an international AI hub in Montreal.[6]

While challenges are certainly part of the future of AI in Montreal and elsewhere, there is no doubt that a new generation is coming and that the story is far from over. Therefore, tracing the formation of this sequence of commons highlight the simultaneous dynamics of the main dimensions of the commons.

4.2. Findings from network analysis: the bottom-up approach and relevant mechanisms

Following the literature on the emergence of innovation ecosystems, we build three hypotheses on the mechanisms underlying the emergence of innovation ecosystems and conducted network analyses to test them. In our first hypothesis, we argue that commoners facilitate the connection between actors, leading to the orchestration of social commons and contributing to the emergence of innovation ecosystems. To test this hypothesis, we draw a network model of the ecosystem using the binary matrix illustrating the connection between actors. In Figure 3, we present a network model of the ecosystem by identifying the linkages between commoners and the organization they are directly or indirectly connected. It is evident from the diagram that there are linkages

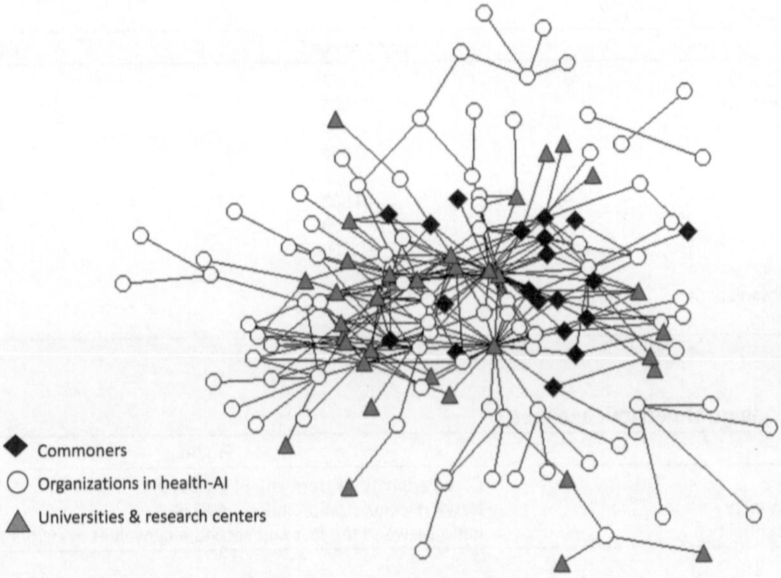

Figure 3. The network diagram of the AI ecosystem in Montreal.

between heterogeneous actors, and there is no specific core or core of actors in the ecosystem.

From the different shapes used to represent commoners and organizations, we observe that commoners are indeed the agents connecting not only different commoners but also different organizational actors, universities, and research centres. Due to their connection to multiple organizations and expertise, commoners in the AI ecosystem in Montreal play crucial roles to connect actors throughout the history of the ecosystem. In other words, commoners facilitate the connection between multiple actors in the ecosystem. The findings provide support for our first hypothesis.

Our second hypothesis is about the structure of the ecosystem. Since an innovation ecosystem is defined as a collaborative arrangement where actors from different domains come together to find innovative solutions, we argue that innovation ecosystems have a decentralized structure and that the decentralized structure of the community of actors leads to the orchestration of symbolic commons and contributes to the emergence of innovation ecosystems. Findings from the core-periphery analysis and the closeness centrality analysis provide support for our argument. We presented the descriptive statistics in Table 3 and findings in Table 4. First, we conducted the core-periphery analysis to investigate whether there is a core in the AI ecosystem in Montreal. The findings of the core-periphery analysis, presented in Table 4, indicate that there is not much distance between the core and the periphery in the AI ecosystem in Montreal (the core/periphery fit = 0.3099) providing support to our second hypothesis. A deeper look at the finding reveals the presence of commoners at the core of the ecosystem, which provides further support to our first hypothesis.

Next, we conducted the closeness centrality analysis to understand the structure of the ecosystem. Closeness centrality is a measure of centrality in a network and indicates how close a network is to all other nodes in the network. The result, the network centralization

Table 3. Descriptive statistics.

Description	Eigenvector 1	Eigenvector 2
Mean	0.053	7.531
Standard deviation	0.066	9.336
Sum	7.402	1046.855
Variance	0.004	87.164
SSQ	1.000	20000.002
MCSSQ	0.606	12115.790
Euc Norm	1.000	141.421
Minimum	−0.000	−0.000
Maximum	0.369	52.250
Number of observations	139.000	139.000
Number missing	0.000	0.000

Table 4. Findings of network analyses.

Analysis	Findings
Core/periphery	Core/Periphery fit (correlation) = 0.3099
Closeness centrality	Network centralization index = 49.23%
Eigenvector centrality	Ratio between the first and second eigenvalues = 6.906/4.216 = 1.424

index = 49.23%, indicates the decentralized structure of the ecosystem (Table 4). In addition, we conducted the eigenvector centrality analysis to confirm the robustness of our finding in the closeness centrality analysis. We found the ratio between the first and second eigenvalues is less than 1.5, confirming our earlier finding. Thereby, our second hypothesis is supported.

Our third hypothesis is about the diversity of actors and overlapping boundaries in innovation ecosystems, leading to the orchestration of knowledge commons and emergence of innovation ecosystems. Analyzing the data, we found evidence to support the hypothesis. We observed that actors in the AI ecosystem come from diverse backgrounds and overlapping boundaries, including private companies, public authorities, major hospitals, universities, research centres, and other support organizations. Many of the actors in the AI ecosystem in Montreal are private companies, and a deeper look into the data shows that over 90% are startups. Further analysis of the data provides us with a detailed understanding of overlapping boundaries in the ecosystem. We found that a commoner can be part of different organizations, and more than one commoner can be connected to one organization. For example, Yoshua Bengio is connected to the University of Montreal and McGill University as well as to MILA and IVADO.

We present the details of the connection of commoners to different organizations in Table 5. One important point to note here is that although there are some public actors in the AI ecosystem in Montreal (Figure 2), none of the commoners we identified is a public actor (Table 5). This finding further supports the bottom-up approach in the emergence of innovation ecosystems.

5. Discussion

In this study, we investigated the AI ecosystem in Montreal to obtain a detailed understanding of the mechanisms underlying the emergence of innovation ecosystems. We identified that the mechanism underlying the emergence of innovation ecosystems in

Table 5. Overlapping boundaries and diversity of actors in the AI ecosystem in Montreal.

Organization Commoners	Support	University	Private	Research	Hospital
C-1		2	9		
C-2		2	7		
C-3		2	6	1	
C-4	1	1	1	3	
C-5		2	2	3	
C-6		2	1	3	
C-7	2	2	1		1
C-8		3			2
C-9		2	3		
C-10		2		3	
C-11		2		3	
C-12		1	1	2	
C-13		1		2	
C-14		1		1	1
C-15		1	1	1	
C-16		1			2
C-17			3		
C-18			1	1	
C-19		1		1	

Montreal is the articulation of innovation commons by commoners. Based on the actions of commoners in different steps of the development of the ecosystem, we identified three sub-mechanisms: orchestration of social commons, orchestration of a symbolic commons, and orchestration of knowledge commons. At first, commoners orchestrate social commons by connecting actors from diverse backgrounds. Next, the decentralized structure of the community of actors leads to the orchestration of symbolic commons. Finally, the orchestration of knowledge commons occurs due to the collaboration between actors across boundaries and their attendant knowledge. Below, we discuss the implications of our findings.

To begin with, our findings highlight that the bottom-up initiatives of commoners are at the origin of the formation of an innovation ecosystem. Commoners play crucial roles in the formation of rich and vibrant innovation ecosystems that have significant impacts on society by enabling knowledge to be progressively revealed, enhanced, nurtured, interpreted, and enacted collectively (Allen and Potts 2016; Cohendet 2022; Helfrich and Haas 2009). Individuals in society can develop private collective actions and solutions to address social dilemmas that may act comparatively well with the hazards of uncertainty as compared to firms, markets, and states (Allen and Potts 2016). For example, collaborative spaces around the world associated with major innovations in society were not issued from a top-down decision from a public authority or a private organization but were the result of a bottom-up initiative carried out by an informal group of passionate individuals (Cohendet 2022). The interplay between actors at different levels plays important role in industrial restructuring and development (Doloreux and Turkina 2021; Isaksen et al. 2019).

The commoners jointly manage the use, access, preservation, and development of the common resource pool including natural resources, culture, knowledge, and innovation (Allen and Potts 2016; Bettiol and Sedita 2011; Ostrom 1990, 2009). The collective efforts of the individuals result in a rich and diverse resource pool (Helfrich and Haas 2009; Willis 2012). The dynamics of the development of innovation in a domain result from heterogeneous interdependencies between different stakeholders including informal groups and communities (Adner and Kapoor 2010; Iansiti and Levien 2004; Smorodinskaya et al. 2017; Yaghmaie and Vanhaverbeke 2019). The decentralized structure of the community of actors is, therefore, required to develop a shared understanding and symbolic commons to work collectively. The network of relationships in innovation ecosystems creates opportunities for the commoners to exchange knowledge and orchestrate knowledge commons. It is important to identify the knowledge and information among a pool of resources and the community that manages such resources to understand the process of innovation and support the development of innovation ecosystems.

Next, our study contributes to the debate on the top-down vs. bottom-up approach to ecosystems. Scholars arguing in favour of the bottom-up approach suggest that the bottom-up perspective provides a detailed understanding to define the challenges at the starting period of an ecosystem, as self-organizing mechanisms evolve slowly from the actions and reactions of participants (Lenkenhoff et al. 2018). For example, Majava et al. (2016) demonstrate the bottom-up nature of San Diego's health and life sciences ecosystem and find that the top-down approach may not be successful in fostering ecosystems. The ideas, knowledge, and information available in society are usually gathered in an open innovation environment through a bottom-up approach which is required for an innovation ecosystem to be successful (Gifford, McKelvey, and Saemundsson 2021; Panetti et al. 2020). The bottom-up creation of synergies and cooperative mechanisms between local actors often makes an innovation ecosystem successful (Panetti et al. 2020). Our observation of the bottom-up approaches in the emergence of the AI innovation ecosystem in Montreal sheds light on the challenges in innovation ecosystems at the beginning and during evolution.

Ecosystems may require the combination of top-down exploration of policy alternatives by policymakers and bottom-up knowledge-intensive entrepreneurial activity to progress toward sustainable development (Colombo et al. 2019; Gifford, McKelvey, and Saemundsson 2021; Lenkenhoff et al. 2018). For example, a combination of the top-down exploration of policy alternatives by policymakers and the bottom-up knowledge-intensive entrepreneurial activity is necessary for sustainable development (Gifford, McKelvey, and Saemundsson 2021). To demonstrate the variance in digital business ecosystem development, Lenkenhoff et al. (2018) illustrate the top-down approach by studying a big company and the bottom-up approach by studying an SME. Policies do play roles in the evolution of an ecosystem, but the roles are different in different ecosystems and played at different times. For example, although the federal and provincial governments recently invested heavily in AI in Montréal, public financing was not at the origin of the emergence of AI activities in the city. In Montreal, the bottom-up initiatives of commoners led to the emergence of the ecosystem of innovation in AI; the top-down support of the AI ecosystem has come mainly during phase III of the emergence. The openness of policymakers to develop bottom-up and place-based digital initiatives is important for the transition to the new digital era (Hervás-Oliver 2021).

Finally, our findings have important implications for understanding the importance of emerging technologies and the digitalization of industries as well as for the development of firms, districts, clusters, cities, regions, and innovation to cope with digital transformations. Emerging technologies such as AI are rapidly changing the status quo in many industries (Hervás-Oliver 2021). For example, in industry 4.0, the logic of the relationship between companies moves from a simple chain to a network, and the capacity of companies to create and capture value would depend on the capability of building and maintaining networks (Götz and Jankowska 2017). Therefore, a successful transformation towards industry 4.0 requires mutual close cooperation between actors and shared norms (Götz and Jankowska 2017). The new industrial revolution and digital transformation will affect not only the high-tech industrial sectors but the whole society. With the emergence of an innovation paradigm, the perspective on innovation processes, their integration into societal structures and processes as well as the object of innovation all are changing (Domanski, Howaldt, and Kaletka 2020). For example, cultural organizations and creative industries may face significant challenges due to the emergence and rapid adoption of digital technologies (Lazzeretti et al. 2022). Industrial clusters and ecosystems facilitate the flow of knowledge and technologies among firms in a location and are highly important for accelerating regional economic performance and innovation (Boix and Trullén 2007; Feldman and Audretsch 1999).

To sum up, it is important to study the relationship between digital transformation and the changes in relevant industries to understand the new dynamics and challenges generated by digital transformation (Lazzeretti et al. 2022). This paper provides new insights into the dynamics and mechanisms contributing to the emergence of innovation ecosystems in a region. Innovation is a fundamental source of wealth generation within an economy (Jackson 2011). Our findings contribute to our understanding of how to facilitate the creation of a successful ecosystem of innovation and formulate favourable innovation policies. For example, contexts along with norms and behaviours shared by a specific group of individuals play crucial roles in the evolution and development of ecosystems (Colombo et al. 2019). By elucidating the role of individuals in the emergence of innovation ecosystems, we provide a reference point for ecosystem analysis.

6. Conclusion

In this study, we investigated the AI ecosystem in Montreal to obtain a detailed understanding of the mechanisms underlying the emergence of innovation ecosystems and find that the mechanism underlying the emergence of innovation ecosystems in Montreal is the articulation of innovation commons by commoners. Based on the actions of commoners in different steps of the development of the ecosystem, we identified three sub-mechanisms: orchestration of social common, orchestration of symbolic common, and orchestration of knowledge commons. At first, commoners orchestrate social commons by connecting actors from diverse backgrounds. Next, the decentralized structure of the community of actors leads to the orchestration of symbolic commons. Finally, the orchestration of knowledge commons occurs due to the collaboration between actors across boundaries and their attendant knowledge.

Given that collaborations between actors from diverse backgrounds have a positive effect on innovation (Innocenti et al. 2022; Nieth et al. 2018), studying the relationship between digital transformation and the changes in relevant industries can help understand the new dynamics and changes generated by digital transformation (Lazzeretti et al. 2022). For example, participants in an innovation ecosystem might decide to leave the ecosystem if the environment is no longer favourable or more relevant innovations are produced elsewhere (Asplund et al. 2021). Our findings indicate the relevance of a systemic and multi-actor perspective on the emergence of innovation ecosystems that incorporates non-firm actors, the relationships between individual and organizational actors, and a shared understanding and vision (Bettiol and Sedita 2011; Hassink, Isaksen, and Trippl 2019; Jolly, Grillitsch, and Hansen 2020). Therefore, the knowledge of various actors needs to be identified and combined in novel ways to define regional capabilities and cope with challenges in the digital era (Boix and Trullén 2007; Nieth et al. 2018).

We acknowledge that ecosystems may require a combination of top-down and bottom-up initiatives. The top-down approach makes sense when the necessary information is available at the top level, and there exists a supportive system to communicate strategies and plans from top to down (Colombo et al. 2019). For example, potential participants of a digital business ecosystem need to establish mechanisms similar to those of natural ecosystems (Lenkenhoff et al. 2018). Policymakers and regulatory bodies might play the central role at some points in an ecosystem, but their role might not be as important at the beginning – while innovation ecosystems emerge. We also acknowledge that our study is not free of limitations, mainly, because of the aspect that we did not compare the dynamics of AI in Montréal to that of other innovative territories. Another aspect is that in the case of Montreal, we concentrate on the successful application of AI in the domain of health. In Montreal, the application of AI in health is particularly strong, but it is important to investigate the deployment of AI in other domains including manufacturing, energy, mobility, or entertainment.

To sum up, our findings provide a detailed understanding of the initiatives of commoners and the mechanisms underlying the emergence of innovation ecosystems as well as offering insights to deal with the rising challenges of digital transformation. The perspective adopted in this contribution opens avenues for future research to understand the dynamics of ecosystems in different domains.

Notes

1. All information used to describe and analyze the AI ecosystem in Montreal comes from the sources mentioned in Appendix A.
2. All information used to describe and analyze the AI ecosystem in Montreal come from the sources mentioned in Appendix A.
3. https://www.montrealinternational.com/en/publications/greater-montreal-an-artificial-intelligence-hub/.
4. https://business-map.montrealinternational.com/en/map/?companysearch=&chk_sector%5B%5D=3&chk_sector%5B%5D=12&chk_sector%5B%5D=13&chk_sector%5B%5D=30&chk_sector%5B%5D=15&chk_sector%5B%5D=11&chk_sector%5B%5D=10&chk_sector%5B%5D=29&chk_sector%5B%5D=28.
5. https://fortune.com/2019/04/02/eye-on-ai-godfathers-deep-learning/.
6. https://www.montrealinternational.com/en/news/the-global-partnership-on-artificial-intelligence-officially-launched/.

Disclosure statement

No potential conflict of interest was reported by the author(s).

ORCID

Ekaterina Turkina http://orcid.org/0000-0002-0549-8237
Patrick Cohendet http://orcid.org/0000-0002-2697-6963

References

Adner, R. 2006. "Match Your Innovation Strategy to Your Innovation Ecosystem." *Harvard Business Review* 84 (4): 98.

Adner, R., and R. Kapoor. 2010. "Value Creation in Innovation Ecosystems: How the Structure of Technological Interdependence Affects Firm Performance in New Technology Generations." *Strategic Management Journal* 31 (3): 306–333. doi:10.1002/smj.821.

Allen, D., and J. Potts. 2016. "The Origin of the Entrepreneur and the Role of the Innovation Commons." Available at SSRN 2867850.

Amin, A., and J. Roberts. 2008. "Knowing in Action: Beyond Communities of Practice." *Research Policy* 37 (2): 353–369. doi:10.1016/j.respol.2007.11.003.

Arena, M., R. Cross, J. Sims, and M. Uhl-Bien. 2017. "How to Catalyze Innovation in Your Organization." *MIT Sloan Management Review* 58 (4): 38–48.

Ashton, W. 2009. "The Structure, Function, and Evolution of a Regional Industrial Ecosystem." *Journal of Industrial Ecology* 13 (2): 228–246. doi:10.1111/j.1530-9290.2009.00111.x.

Asplund, F., J. Bjork, M. Magnusson, and A. Patrick. 2021. "The Genesis of Public-Private Innovation Ecosystems: Bias and Challenges." *Technological Forecasting and Social Change* 162: Article 120378. doi:10.1016/j.techfore.2020.120378.

Autio, E., and L. Thomas. 2014. "Innovation Ecosystems." In *The Oxford Handbook of Innovation Management*, edited by Mark Dodgson, David M. Gann, and Nelson Phillips, 204–288. doi:10.1093/oxfordhb/9780199694945.013.01.

Bathelt, H., A. Malmberg, and P. Maskell. 2004. "Clusters and Knowledge: Local Buzz, Global Pipelines and the Process of Knowledge Creation." *Progress in Human Geography* 28 (1): 31–56. doi:10.1191/0309132504ph469oa.

Bettiol, M., and S. R. Sedita. 2011. "The Role of Community of Practice in Developing Creative Industry Projects." *International Journal of Project Management* 29 (4): 468–479. doi: 10.1016/j.ijproman.2011.01.007.

Boix, R., and J. Trullén. 2007. "Knowledge, Networks of Cities and Growth in Regional Urban Systems." *Papers in Regional Science* 86 (4): 551–574. doi:10.1111/j.1435-5957.2007.00139.x.

Boréus, K., and G. Bergström. 2017. *Analyzing Text and Discourse: Eight Approaches for the Social Sciences*. London: Sage.

Borgatti, S., M. Everett, and J. Johnson. 2018. *Analyzing Social Networks*. Los Angeles: Sage.

Bowers, C. 2004. "Revitalizing the Commons or an Individualized Approach to Planetary Citizenship: The Choice Before Us." *Educational Studies* 36 (1): 45–45. doi:10.1207/s15326993es3601_5.

Cohendet, P. 2022. "Architectures of the Commons: Collaborative Spaces and Innovation." *ZFW – Advances in Economic Geography* 66 (1): 36–48. doi:10.1515/zfw-2022-0008.

Cole, A., and D. Barberá-Tomás. 2014. "Negotiating Conventions and Creating Community: The Case of Cartoon and European Animation." *Journal of Economic Geography* 14 (5): 973–993. doi:10.1093/jeg/lbu025.

Colombo, M., G. Dagnino, E. Lehmann, and M. Salmador. 2019. "The Governance of Entrepreneurial Ecosystems." *Small Business Economics* 52 (2): 419–428. doi:10.1007/s11187-017-9952-9.

Cowan, R., P. David, and D. Foray. 2000. "The Explicit Economics of Knowledge Codification and Tacitness." *Industrial and Corporate Change* 9 (2): 211–253. doi:10.1093/icc/9.2.211.

DeLanda, M. 2016. *Assemblage Theory*. Edinburgh: Edinburgh University Press. doi:10.1515/9781474413640.

DeLanda, M. 2019. *A New Philosophy of Society: Assemblage Theory and Social Complexity*. London: Bloomsbury Publishing.

de Vasconcelos Gomes, L., A. Facin, M. Salerno, and R. Ikenami, 2018. "Unpacking the Innovation Ecosystem Construct: Evolution, Gaps and Trends." *Technological Forecasting and Social Change* 136: 30–48. doi:10.1016/j.techfore.2016.11.009.

Doloreux, D., and E. Turkina. 2021. "New Path Creation in the Artificial Intelligence Industry: Regional Preconditions, New Actors and Their Collective Actions, and Policies." *Regional Studies* 55 (10-11): 1751–1763. doi:10.1080/00343404.2021.1919610.

Domanski, D., J. Howaldt, and C. Kaletka. 2020. "A Comprehensive Concept of Social Innovation and Its Implications for the Local Context – On the Growing Importance of Social Innovation Ecosystems and Infrastructures." *European Planning Studies* 28 (3): 454–474. doi:10.1080/09654313.2019.1639397.

Faissal Bassis, N., and F. Armellini. 2018. "Systems of Innovation and Innovation Ecosystems: A Literature Review in Search of Complementarities." *Journal of Evolutionary Economics* 28 (5): 1053–1080. doi:10.1007/s00191-018-0600-6.

Feldman, M. 2005. The Locational Dynamics of the U.S. Biotech Industry: Knowledge Externalities and the Anchor Hypothesis. In *Research and Technological Innovation*, edited by A. Quadrio Curzio and M. Fortis, 201–224. doi:10.1007/3-7908-1658-2_9.

Feldman, M., and D. Audretsch. 1999. "Innovation in Cities: Science-Based Diversity, Specialization and Localized Competition." *European Economic Review* 43 (2): 409–429. doi:10.1016/S0014-2921(98)00047-6.

Fu, P. 2021. "From Bottom-Up to Top-Down: Governance, Institutionalisation, and Innovation in Chinese Makerspaces." *Technology Analysis & Strategic Management* 33 (10): 1226–1241. doi:10.1080/09537325.2021.1950680.

Gifford, E., M. McKelvey, and R. Saemundsson. 2021. "The Evolution of Knowledge-Intensive Innovation Ecosystems: Co-Evolving Entrepreneurial Activity and Innovation Policy in the West Swedish Maritime System." *Industry and Innovation* 28 (5): 651–676. doi:10.1080/13662716.2020.1856047.

Götz, M., and B. Jankowska. 2017. "Clusters and Industry 4.0 – Do They Fit Together?" *European Planning Studies* 25 (9): 1633–1653. doi:10.1080/09654313.2017.1327037.

Grandadam, D., P. Cohendet, and R. Suire. 2022. "Building and Nurturing Grassroots Innovation: A Policy Framework Based on the Local Commons." *European Planning Studies* 30 (8): 1577–1595. doi:10.1080/09654313.2021.1998385.

Granstrand, O., and M. Holgersson. 2020. "Innovation Ecosystems: A Conceptual Review and a New Definition." *Technovation* 90-91 (12): Article 102098. doi:10.1016/j.technovation.2019.102098.

Hassink, R., A. Isaksen, and M. Trippl. 2019. "Towards a Comprehensive Understanding of New Regional Industrial Path Development." *Regional Studies* 53 (11): 1636–1645. doi:10.1080/00343404.2019.1566704

Helfrich, S., and J. Haas. 2009. "The Commons: A New Narrative for Our Times." In *Genes, Bytes and Emissions: To Whom Does the World Belong*, edited by Silke Helfrich, 1–15. Heinrich Böll Foundation.

Hervás-Oliver, J. 2021. "Industry 4.0 in Industrial Districts: Regional Innovation Policy for the Toy Valley District in Spain." *Regional Studies* 55 (10-11): 1775–1786. doi:10.1080/00343404.2021.1939861.

Iansiti, M., and R. Levien. 2004. *The Keystone Advantage: What the New Dynamics of Business Ecosystems Mean for Strategy, Innovation, and Sustainability*. Boston, Mass: Harvard Business School Press.

Innocenti, N., F. Capone, L. Lazzeretti, and S. Petralia. 2022. "The Role of Inventors' Networks and Variety for Breakthrough Inventions." *Papers in Regional Science* 101 (1): 37–57. doi:10.1111/pirs.12640.

Isaksen, A., S. Jakobsen, R. Njøs, and R. Normann. 2019. "Regional Industrial Restructuring Resulting from Individual and System Agency." *Innovation: The European Journal of Social Science Research* 32 (1): 48–65. doi:10.1080/13511610.2018.1496322.

Jackson, D. 2011. "What Is an Innovation Ecosystem." *National Science Foundation* 1 (2): 1–13.

Jacobides, M., C. Cennamo, and A. Gawer. 2018. "Towards a Theory of Ecosystems." *Strategic Management Journal* 39 (8): 2255–2276. doi:10.1002/smj.2904.

Jolly, S., M. Grillitsch, and T. Hansen. 2020. "Agency and Actors in Regional Industrial Path Development. A Framework and Longitudinal Analysis." *Geoforum; Journal of Physical, Human, and Regional Geosciences* 111: 176–188. doi:10.1016/j.geoforum.2020.02.013.

Karakas, F. 2009. "Welcome to World 2.0: The New Digital Ecosystem." *Journal of Business Strategy* 30 (4): 23–30. doi:10.1108/02756660910972622

Kirwan, S., Dawney, L., and Brigstocke, J, eds. 2015. *Space, Power and the Commons: The struggle for alternative futures.* 1st ed. London: Routledge. doi:10.4324/9781315731995

Lacouture, A., E. Breton, A. Guichard, and V. Ridde. 2015. "The Concept of Mechanism from a Realist Approach: A Scoping Review to Facilitate Its Operationalization in Public Health Program Evaluation." *Implementation Science* 10 (1): 1–10. doi:10.1186/s13012-015-0345-7.

Lazzeretti, L., S. Oliva, N. Innocenti, and F. Capone. 2022. "Rethinking Culture and Creativity in the Digital Transformation." *European Planning Studies*, 1–9.

Lenkenhoff, K., U. Wilkens, M. Zheng, T. Süße, B. Kuhlenkötter, and X. Ming. 2018. "Key Challenges of Digital Business Ecosystem Development and How to Cope with Them." *Procedia Cirp* 73: 167–172. doi:10.1016/j.procir.2018.04.082.

Longhi, C. 2005. A French revolution: technology management in the aerospace industry. The case of Toulouse. *International Journal of Technology Management* 29 (3-4): 194–215. doi:10.1504/IJTM.2005.005996.

Majava, J., P. Leviäkangas, T. Kinnunen, P. Kess, and D. Foit. 2016. "Spatial Health and Life Sciences Business Ecosystem: A Case Study of San Diego." *European Journal of Innovation Management* 19 (1): 26–46. doi:10.1108/EJIM-01-2015-0003.

Moore, J. 1993. "Predators and Prey: A New Ecology of Competition." *Harvard Business Review* 71 (3).

Ngongoni, C., S. Grobbelaar, and C. Schutte. 2022. "Making Sense of the Unknown: Using Change Attractors to Explain Innovation Ecosystem Emergence." *Systemic Practice and Action Research* 35: 227–252. doi:10.1007/s11213-021-09564-x.

Nieth, L., P. Benneworth, D. Charles, L. Fonseca, C. Rodrigues, M. Salomaa, and M. Stienstra. 2018. "Embedding Entrepreneurial Regional Innovation Ecosystems: Reflecting on the Role of Effectual Entrepreneurial Discovery Processes." *European Planning Studies* 26 (11): 2147–2166. doi:10.1080/09654313.2018.1530144.

Oh, D., F. Phillips, S. Park, and E. Lee. 2016. "Innovation Ecosystems: A Critical Examination." *Technovation* 54: 1–6. doi:10.1016/j.technovation.2016.02.004.

Ostrom, E. 1990. *Governing the Commons: The Evolution of Institutions for Collective Action.* Cambridge, UK: Cambridge University Press.

Ostrom, E. 2009. "What Is Social Capital." In *Social Capital: Reaching out, Reaching in*, edited by V. Bartkus, and J. Davis, 17–38. Cheltenham: Edward Elgar.

Paasi, J., H. Wiman, T. Apilo, and K. Valkokari. 2022. "Modeling the Dynamics of Innovation Ecosystems." *International Journal of Innovation Studies.* doi:10.1016/j.ijis.2022.12.002.

Panetti, E., A. Parmentola, M. Ferretti, and E. Reynolds. 2020. "Exploring the Relational Dimension in a Smart Innovation Ecosystem: A Comprehensive Framework to Define the Network Structure and the Network Portfolio." *The Journal of Technology Transfer* 45 (6): 1775–1796. doi:10.1007/s10961-019-09735-y.

Radziwon, A., M. Bogers, and A. Bilberg. 2017. "Creating and Capturing Value in a Regional Innovation Ecosystem: A Study of How Manufacturing SMEs Develop Collaborative Solutions." *International Journal of Technology Management* 75 (1-4): 73–96. doi:10.1504/IJTM.2017.085694.

Ritala, P., and R. Gustafsson. 2018. "Q&A. Innovation and Entrepreneurial Ecosystem Research: Where Are We Now and How Do We Move Forward?" *Technology Innovation Management Review* 8 (7): 52–57. doi:10.22215/timreview/1171.

Romero, D., and A. Molina. 2011. "Collaborative Networked Organisations and Customer Communities: Value Co-Creation and Co-Innovation in the Networking Era." *Production Planning and Control* 22 (5-6): 447–472. doi: 10.1080/09537287.2010.536619.

Sætre, A., and A. Van de Ven. 2021. "Generating Theory by Abduction." *Academy of Management Review* 46 (4): 684–701. doi:10.5465/amr.2019.0233.

Saxenian, A. 2018. Inside-out: regional networks and industrial adaptation in Silicon Valley and Route 128. In *The sociology of economic life*, edited by Mark Granovetter, 357–374. Routledge. doi:10.4324/9780429494338.

Simmonds, H., A. Gazley, V. Kaartemo, M. Renton, and V. Hooper. 2021. "Mechanisms of Service Ecosystem Emergence: Exploring the Case of Public Sector Digital Transformation." *Journal of Business Research* 137: 100–115. doi:10.1016/j.jbusres.2021.08.008.

Smith, A., and A. Stirling. 2018. "Innovation, Sustainability and Democracy: An Analysis of Grassroots Contributions." *Journal of Self-Governance and Management Economics* 6 (1): 64–97. doi:10.22381/JSME6120183.

Smorodinskaya, N., M. Russell, D. Katukov, and K. Still. 2017. "Innovation Ecosystems vs. Innovation Systems in Terms of Collaboration and Co-Creation of Value." In Proceedings of the 50th Hawaii International Conference on System Sciences, January.

Sun, S., Y. Zhang, Y. Cao, J. Dong, and J. Cantwell. 2019. "Enriching Innovation Ecosystems: The Role of Government in a University Science Park." *Global Transitions* 1: 104–119. doi:10.1016/j.glt.2019.05.002.

Thomas, E., K. Faccin, and B. Asheim. 2021. "Universities as Orchestrators of the Development of Regional Innovation Ecosystems in Emerging Economies." *Growth and Change* 52 (2): 770–789. doi:10.1111/grow.12442.

Tsujimoto, M., Y. Kajikawa, J. Tomita, and Y. Matsumoto. 2018. "A Review of the Ecosystem Concept — Towards Coherent Ecosystem Design." *Technological Forecasting and Social Change* 136: 49–58. doi:10.1016/j.techfore.2017.06.032.

Valkokari, K. 2015. "Business, Innovation, and Knowledge Ecosystems: How They Differ and How to Survive and Thrive Within Them." *Technology Innovation Management Review* 5: 17–24. doi:10.22215/timreview/919.

Von Krogh, G., and N. Geilinger. 2014. "Knowledge Creation in the Eco-System: Research Imperatives." *European Management Journal* 32 (1): 155–163. doi:10.1016/j.emj.2013.04.002.

Willis, P. 2012. "Engaging Communities: Ostrom's Economic Commons, Social Capital and Public Relations." *Public Relations Review* 38 (1): 116–122. doi:10.1016/j.pubrev.2011.08.016.

Yaghmaie, P., and W. Vanhaverbeke. 2019. "Identifying and Describing Constituents of Innovation Ecosystems: A Systematic Review of the Literature." *Euromed Journal of Business* 15 (3): 283–314. doi:10.1108/EMJB-03-2019-0042.

Zimmermann, J. 2020. *Les communs : des jardins partagés à Wikipedia*. Paris: Éditions Libre et Solidaire. doi:10.4000/terminal.7334.

Appendix A. Data sources

https://www.thecanadianencyclopedia.ca/en/article/artificial-intelligence
https://www.forbes.com/sites/peterhigh/2017/11/06/why-montreal-has-emerged-as-an-artificial-intelligence-powerhouse/?sh=fcaa2a823bd4
http://www.canada.ai/timeline
http://www.montrealintechnology.com/ai-leaders-in-montreal-launch-ai-startup-factory/
https://medium.com/real-ventures/canadas-artificial-intelligence-ecosystem-4798b0517016
https://ormuco.com/blog/canada-pioneers-ai-research-policy#:~:text=Most%20prominent%20of%20Montrealais%20AI,co%2Dfounder%20of%20Element%20AI
https://www.forbes.com/sites/peterhigh/2017/11/06/why-montreal-has-emerged-as-an-artificial-intelligence-powerhouse/?sh=7fd4f97a23bd
https://www.concordia.ca/jmsb/news/magazine/issue-2/the-reality-of-ai-in-montreal.html
https://www.montrealinternational.com/en/news/lintelligence-artificielle-au-service-de-la-sante/
https://www.montrealinternational.com/en/news/life-sciences-and-health-technology-the-compelling-attraction-of-montreal/
https://ici.radio-canada.ca/nouvelle/1926656/yoshua-bengio-palmares-scientifiques-plus-influent-monde-ia
https://www.montrealinternational.com/en/publications/greater-montreal-an-artificial-intelligence-hub/
https://www.canada.ca/en/innovation-science-economic-development/news/2019/09/government-of-canada-and-government-of-quebec-announce-the-creation-of-an-international-centre-of-expertise-in-montreal-for-the-advancement-of-arti.html
https://business-map.montrealinternational.com/en/map/?companysearch=&chk_sector%5B%5D=3&chk_sector%5B%5D=12&chk_sector%5B%5D=13&chk_sector%5B%5D=30&chk_sector%5B%5D=15&chk_sector%5B%5D=11&chk_sector%5B%5D=10&chk_sector%5B%5D=29&chk_sector%5B%5D=28

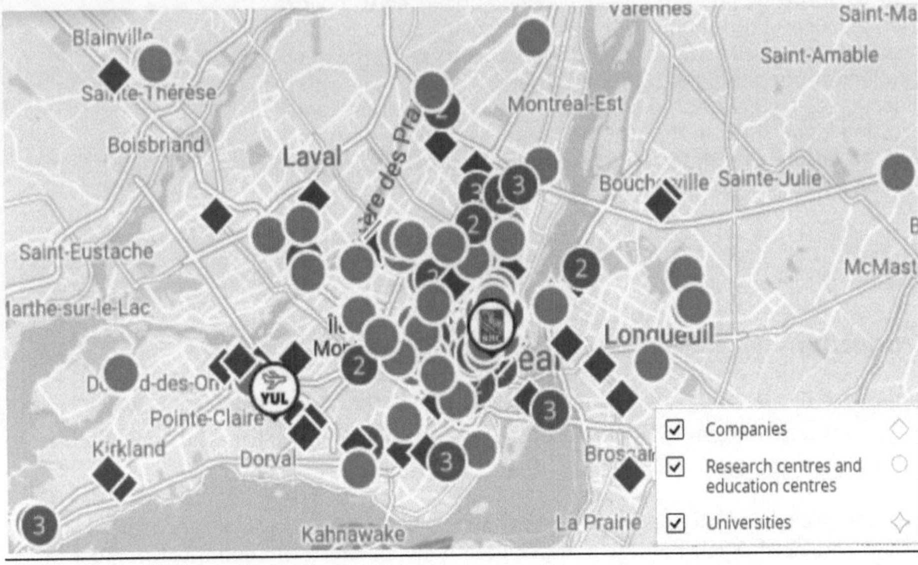

How real will the metaverse be? Exploring the spatial impact of virtual worlds

Pier Luigi Parcu, Maria Alessandra Rossi, Niccolò Innocenti and Chiara Carrozza

ABSTRACT
In this paper, we perform a preliminary analysis of the technologies, firms and industries that may be affected by the possible futures of the metaverse, attempting to derive some hypotheses on the spatial effects of this process. We distinguish between two possible evolutive scenarios – the 'metaverse shaped by reality view' and the 'metaverse shaping reality view' – and factors affecting them, deriving implications for public policy planning. The first scenario presents relatively traditional core policy challenges: ensuring homogeneous availability of network infrastructures as well as the skills indispensable to catch the new technological opportunities at the local level, accompanying the reallocation of factors of production associated to disruption and addressing inequalities. In the second, the main challenge is more radical: to ensure that desirable features are incorporated in the emerging virtual worlds from the start.

1. Introduction

The metaverse has been defined in many ways: as a product or a service, as a place, and even as a moment in time.[1] For instance, the author and investor Matthew Ball, in a seminal book on the topic (Ball 2022), described the metaverse as

> a massively scaled and interoperable network of real-time rendered 3D virtual worlds that can be experienced synchronously and persistently by an effectively unlimited number of users with an individual sense of presence, and with continuity of data, such as identity, history, entitlements, objects, communications, and payments.

This definition, along with many others, puts emphasis on the virtual nature of this (set of) technological developments. However, the virtual nature of the metaverse does not exclude the possibility of very concrete spatial economic effects. The technology may potentially have a relevant impact at the local level, influencing both the national dimension as well as the regional and city dimension, modifying the way businesses are conducted and the interactions among people.

The scale of this impact and its concrete nature will depend on whether the metaverse will deliver on its promises. At one extreme, entirely new forms of social interaction may develop, as well as new connections between virtual and physical experiences. This is the optimistic vision for the future of the metaverse, one where the various technologies will converge in supporting the emergence of a new virtual environment where economic concepts and institutions such as *scarcity*, *property* and *markets* take on new meanings. At the other extreme, the metaverse may end up being a 'catch-all' fashionable label to indicate a disparate set of technologies related to virtual experiences, each of which may have a relatively independent evolutionary path, but all substantially replicating existing human activities in an immersive reality. Even in this case, although the vision of a network of virtual worlds populated by avatars would not become central to human experience, the innovations associated with the metaverse may significantly modify our lives, our culture and the way we interact. Almost all sectors that can benefit from a direct, even if virtual, engagement of users may potentially be affected and transformed. What scenario will emerge cannot currently be accurately predicted. The perception of real technical achievements is still uncertain (Dwivedi et al. 2022), investment values in the financial markets are extremely volatile and skepticism on the reliability of the cryptocurrencies associated with the different virtual worlds is also rising (Vidal-Tomás 2022 and 2023). Finally, and importantly, uncertainties on the nature of the use cases beyond gaming and the current relatively low number of users, even at a time when the concept has received significant attention from the public, raise doubts on the viability of the metaverse on a large scale. Nonetheless, in this paper, we adopt a qualitative methodology to conduct a preliminary analysis of the technologies, firms and industries that may be affected by the possible futures of the metaverse, attempting to derive some hypotheses on the spatial effects of this process. Despite its inevitable limitations, we believe such an analysis is opportune in the attempt to separate hype from reality and to plan appropriate policy responses.

Planning will need to evolve in different directions according to whether it will be current reality to shape the metaverse – i.e. the metaverse will remain a label for a variety of technologies that complement existing human activities and sectors through virtualization – or the metaverse will enable new realities through new forms of social interaction. In the first case, the main challenge will be a traditional one: to enact policies that support territories in gauging the innovation and development opportunities opened by frontier innovations and to ensure territorial cohesion by preventing excessively skewed outcomes in terms of the distribution of the benefits. In the second case, there is an additional, more extreme, challenge for planning. The metaverse as a well-identified network of virtual worlds may become an environment where technology will shape human interactions more deeply than it has ever been the case because man-made governance rules are consciously embedded in technology. This takes the idea that the direction of technological evolution is influenced by conscious human choices that some economists have set at the centre of their analysis of inequality (Atkinson 2015) to the next level. In this possible (though perhaps unlikely) future, planning would need to address the fundamental question of how to design the framework within which social interactions will take place in the new virtual space so that desirable features are embedded in the system from the start (Lazzeretti 2023).

The paper is structured as follows. Section 2 examines the hype that currently surrounds the metaverse concept, both among the public and in the academic community. Section 3 starts addressing the issue of the local (country-level) impact of the metaverse by considering the technological readiness of various countries in terms of the outcome of past investments in key technologies related to the production side of the metaverse, as presently captured by patents and scientific publications. Section 4 focuses on the industry-level and adoption-related impact of the metaverse on the local economies by developing three hypotheses, with the associated implications for planning. The focus of this section is on the above-described 'minimalist' view of the metaverse as a set of technologies and related business models, rather than on the fully-fledged network of virtual worlds (the 'metaverse shaped by reality' view). Section 5 speculates on the 'metaverse shaping reality' view of the emergence of a new space of social interaction. It attempts to single out key dimensions of its evolution, again highlighting (even more speculative) implications for planning. Section 6 concludes.

2. The hype around the metaverse and the reality

The year 2022 can be defined, in many ways, the year of the metaverse: interest among the general population has spiked in the months between the end of 2021 and the beginning of 2022, starting at the end of October 2021, when Mark Zuckerberg announced the rebranding from Facebook to Meta. The latter, meant to be an explicit signal of the company's commitment to the metaverse, accelerated the development, the investments in, and the awareness of, the concept. Facebook also bet more than $10 billion on the development of the metaverse in 2021 alone (Kraus et al. 2022). This section provides some quantitative information on the evolution of public, academic and financial attention towards the metaverse (Section 2.1) and summarizes key demand and technological challenges to its development (Section 2.2).

2.1. The fluctuating attention towards the metaverse

Figure 1 presents the search trend on Google for the word metaverse in the last two years, showing that between January/October 2021, searches for the concept were relatively low in number, and have dramatically increased immediately after the rebranding of Facebook, in November 2021/mid-February 2022. The searches subsequently showed a decline, but still settled at a significantly higher value than in October 2021, suggesting

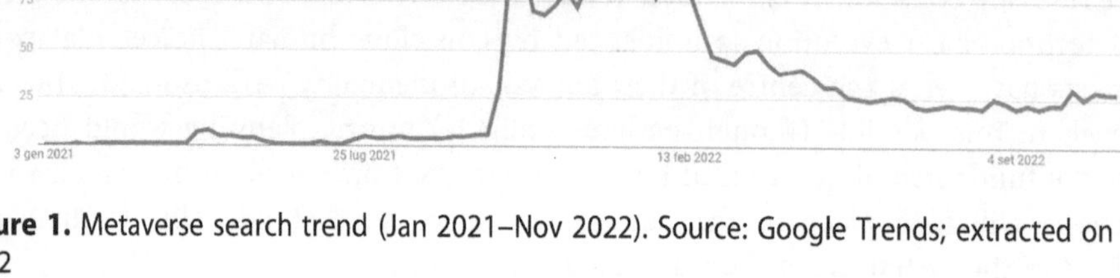

Figure 1. Metaverse search trend (Jan 2021–Nov 2022). Source: Google Trends; extracted on 11-11-2022

that the idea of the metaverse has finally reached wider recognition, beyond technologists and specialized investors.

Attention by investors and the financial community is also increasing considerably. Many influential companies around the globe are investing to build their own visions of the virtual worlds, by also acquiring smaller companies that are considered to be at the frontier of the technology (e.g., Meta has acquired Oculus VR, Microsoft is buying Activision Blizzard). The latter transaction is under severe review by the EU, UK and U.S.A.[2], and by other jurisdictions, showing the attention also of the regulatory authorities. Direct investment by many different companies brings the overall amount of funds well beyond the large investments of Facebook/META.

While academic interest in the metaverse precedes the latest moment of hype just described, even the academic community has shown a remarkable increase of interest in the concept along with the public. Figure 2 shows the rising trend in the yearly number of scientific publications (ISI publications on scientific journals[3]), particularly from 2008, but also the tremendous spike during the last year, with more than 380 papers published in 2022 alone[4] (more than the overall number of publications from 1995 to 2020). Citations for articles with the topic metaverse, another important indicator of scientific attention, also show a value greater than 900 for the year 2022, which is more than nine times the value recorded in the previous year. The bulk of these publications is in the computer science, engineering and telecommunications domains, and therefore concerns mostly technical aspects of the development of the metaverse. More limited, but rapidly growing, is the interest shown by the social sciences, which account for approximately 10%–15% of the publications.

Over the course of the past year, the hype around the metaverse has also patently met reality, raising doubts on the effective likelihood that this potential technological revolution will deliver on its promises. Nowhere have faltering expectations been more visible than in the financial markets. The best-known company that has decided to bet strongly on the metaverse, Meta, has clearly not been rewarded by investors. Meta has recently demonstrated severely negative performances, with a loss in shares value that was

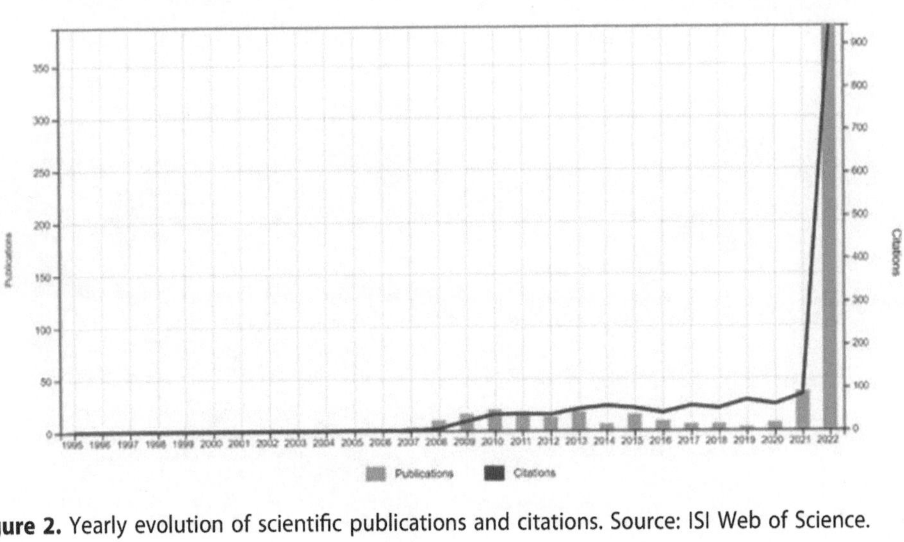

Figure 2. Yearly evolution of scientific publications and citations. Source: ISI Web of Science.

close to 63% during the year 2022, albeit in a context of general retreat of technological titles (Figure 3). Between 2022 and 2023, the company announced that, for the first time in its history, it would be making a staff reduction of approximately 21.000 employees (Financial Times 2023). Also, in March 2023, Meta has informed investors that it will redirect some of its efforts away from the metaverse and towards improving more traditional content recommendation systems through AI applications.

ETF instruments related to the metaverse have also displayed a recent negative performance (Vidal-Tomás 2022). All these financial investments showed a loss greater than 50% in 2022. The largest ETF investing in the metaverse – Roundhill Ball Metaverse ETF[5] – counting on an initial value of Assets Under Management (AUM) of over $850 million, lost approximately 53%. Even if compared to the general decrease in the value of the financial market over the same period, the comparative loss remains about 20% higher. In the last year, other metaverse-focused ETFs, or portfolios, have shown similarly negative performances (e.g. eToro Metaverse Life Smart Portfolio lost 65%; Evolve Metaverse ETF 55% etc).

Financial performance is not necessarily a reliable indicator of the future success of any given technological development, although it does provide some indications on the state of expectations of financial markets and on the ease with which technological bets can be pursued. The next subsection focuses on two other indicators that may provide an explanation of the uncertainties emerging from financial markets, and pertinent information to assess whether there is substance beyond the hype: demand and technological constraints.

2.2. Demand and technological constraints to the evolution of the metaverse

Digital technology has, in the last two decades, made great steps forward and can today enable a more immersive experience: the possibility to easily access the virtual worlds at anytime and from anywhere, thanks to mobile devices, and a deeper bond with real life, for example, by allowing the trading of virtual objects and the use of virtual currencies.

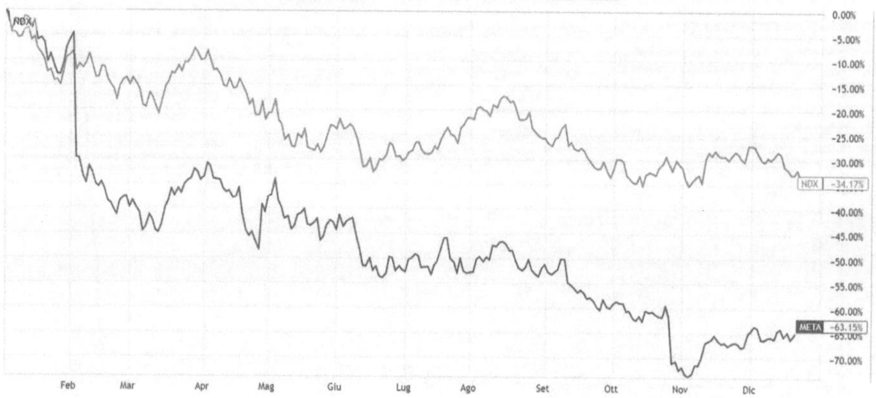

Figure 3. Nasdaq and Meta shares comparison, year 2022. Source: Our elaboration

Thanks to these evolutions, the metaverse, even in its infancy, offers an experience that is quite advanced in comparison to the previous elementary examples of virtual reality (such as Second Life). However, enormous challenges and potential side effects for users today, still partially unknown, remain. To better understand these constraints, it is useful to consider the state of development of the three main components of the metaverse: infrastructure/hardware, software and content (Radoff 2022; Park and Kim 2022).

The infrastructure/hardware layer (fibre networks, WiFi, 5G and beyond, cloud systems, semiconductors) includes the technologies that will make all the other layers of the metaverse possible. Differently from Internet 2.0, this layer needs to include also the new instruments that are necessary to build the human interface with the metaverse.[6] With Internet 2.0, the screen of a pc or of a smartphone was a sufficient interface, while a full metaverse immersive experience requires ad hoc tools that are capable of being intermediaries between humans and the infrastructure. These devices include all the wearable electronics that are in development, such as smart glasses, headsets, gloves and prospectively also deeper forms of man-machine connections. While developing high-performing, convenient, affordable devices presents its challenges, it is in the infrastructure layer that, at the moment, reside the greatest technological hurdles. Fast symmetric bandwidth (allowing users to send as much information as they can receive), low network latency (reducing the time it takes for data to travel from one point to another of the network and back) and overall network speed, all represent crucial features for the metaverse's developments and today's networks cannot meet these needs without significant, possibly radical, changes.

Furthermore, the input collected by the hardware needs to be processed by software to create the models that represent the virtual environment. Although much progress has been made in the past few years in all the software areas related to the development of the metaverse, improvements are still needed to enable a fully-fledged satisfactory virtual user experience (Mozumder et al. 2022). It is possible to single out two main areas of software innovation that may have a decisive impact on the future evolutionary trajectory of the metaverse and therefore on the real impact that it may have at the local level: Decentralization and Spatial Computing. They both do not have clear correspondence to anything in Internet 2.0. Decentralization relates to a vast area of both technological (e.g. distributed ledger technologies, crypto assets and NFTs) and governance innovation that should support achievement of the full potential of the metaverse by assuring a transparent, secure and traceable way to perform transactions and interactions. Spatial computing, in turn, groups a large category of technologies that are related to the access to, and manipulation of, 3D spaces, and to augment the real world with more information and experience.

Finally, the third key component of the metaverse is content. In the 'hype' version, users are expected to create large amounts of multimedia and text content through their avatars, and these entities and their relationships are used to organize events that are combined to form a scenario – something that is at present far from occurring. Nonetheless, content generation probably is the least stringent of the technological constraints because it can be rapidly overcome once a critical mass of users is achieved and indirect network effects start to exert their multiplicative magic. Following a well acknowledged perspective this may be considered as an 'era of ferment' (Anderson and Tushman 1990) that is thus expected to be followed by a period of incremental technical progress that will enable a wider adoption of the metaverse, leading to a subsequent technological discontinuity.

This points to the feedback effects between technological and demand constraints. Indeed, lack of demand for usages (other than gaming) currently appears the other strong constraint to the development of the metaverse. As for any other network/multi-sided platform, the success of the metaverse hinges on its ability to attract a critical mass of users of different types (e.g. end-users and providers of experiences or services). This, however, seems to be very far from occurring. Following the most recent statistics, there are at present more than 400 million monthly unique users, but they are divided among many different metaverses, with a great predominance of the gaming ones (Roblox, with more than 200 million, Minecraft, with 160 million and Fortnite, with 80 million). This also has consequences for the average age of those users, who are largely younger than 18 years old (Metaversed 2022), and who are thus often unable to make transactions, buy assets, and/or spend easily.[7] The prevalence of gaming platforms leaves a few users sparsely scattered among the many different non-gaming platforms. While estimates may vary according to different sources, available data range between the 200.000 monthly users of Horizon World and the 40–50.000 of Sandbox and Decentraland. These, for the moment, appear to be numbers that may allow for limited interaction, and may thus justify the impression that metaverses are mostly empty worlds.

In this connection, one possible demand-side limitation to wider participation to the metaverse may come from the price level of the most advanced instruments needed to allow for a complete and immersive experience (i.e. the 'professional' VR or AR devices). The best devices have priced higher than $1.000 (HTC Vive Focus 3 and Meta Quest Pro), while VR devices that are expected to come from Apple in 2023 will probably be priced in a price range between $2-3.000. These prices appear to be far too high for mass consumption.[8]

3. A look at countries' technological readiness for the production of metaverse-related technologies through patents and scientific publications

To help ground the analysis of the local impact of the metaverse it may be useful to consider two standard indicators of local knowledge/innovation production: patents and scientific publications. Both have well-known limits in terms of their ability to capture innovativeness (Acs, Anselin, and Varga 2002; Burhan, Singh, and Jain 2017; Dernis, Squicciarini, and de Pinho 2015). Despite the limitations, however, a look at patents and publications may provide a first indication of, on one side, the interest shown in different countries towards metaverse-related technologies and, on the other side, of the countries that may be expected to play an active role in the development of these technologies and therefore to have a first mover advantage in the production of metaverse components. This, of course, does not mean that countries that are currently or prospectively not active on the production side will be excluded from participating to the usage side of the metaverse.

In this section, based on the Espacenet database, we will explore the evolution of the patents that are related to the most important of the technologies that are the foundation of the metaverse. Two of them relate to the above-described area of spatial computing –

Augmented Reality (AR) and Virtual Reality (VR) – and the last one is a frontier technology that aims to replace physical hardware (Brain-Computer Interfaces, BCI). A limitation of this method is that a large amount of the metaverse's development is achieved through computer programming improvements and these kinds of innovations normally are not found in the public domain. However, advancements in AR, VR and BCI, considered among the most important of the technologies that may favour the future massive adoption of the metaverse, are normally patented innovations.

The less known and established of the relevant technologies, BCI, shows an overall number of patents that is slightly over 5.300, while AR has just over 136.000, and VR approximately 183.000. By simply using the keyword 'metaverse' we can identify 'only' another 1.540 patents. Running the search again, considering all the technologies together (to avoid double counting) we reach an overall result of 241.567 patents.

Figure 4 shows that this patenting activity displays a significant growth year on year and an acceleration in the past 5 years, confirming the overall growing investors' interest in the metaverse discussed in the previous section: from 1.888 patents in 2009, to 3.096 in 2010, to 9.832 patents in 2015, 19.419 in 2016, until it reached a peak of 38.395 in 2020.

Of course, it is a well-known fact that, in the last twenty years, many kinds of patents have shown important growth. Figure 5 shows that growth of the patents here considered has been higher than average. In the period from 2015 to 2020, the number of patents that are in some way related to the metaverse showed a threefold increase with respect to other technologies, bringing these technologies to grow in percentage from 1,5% to more than 4,5% of the total (Figure 5).

The simple count of patents is, for many reasons, not a refined measure of the development of a technology (Parcu, Innocenti, and Carrozza 2022), however, they undoubtedly provide an acceptable proxy of the extent of the interest different countries show towards the metaverse. Moreover, they may be considered to provide a very rough indication of the availability within a country of knowledge resources that may influence countries' future ability to participate in the development of the relevant technological trajectories. In industries unrelated to the metaverse, patent endowments have been shown to influence the patterns of technological specialization (Belloc and Pagano 2012). Whether this will be the case also in the key metaverse-related technological domains cannot be foreseen at this stage, but it is nonetheless interesting to explore country-level patent endowments.

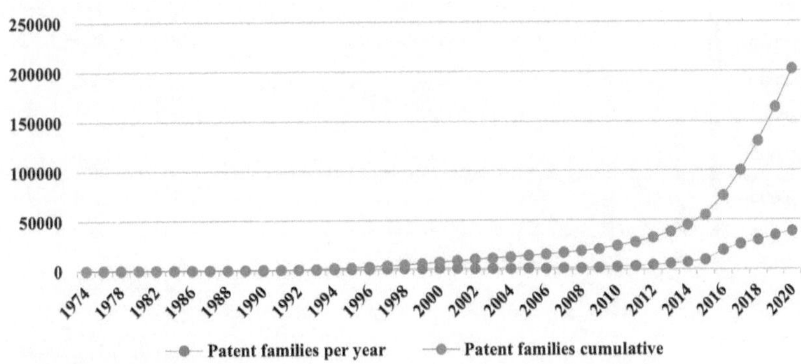

Figure 4. Patent Families AR, VR, BCI, Metaverse. Source: Our elaboration on Espacenet data.

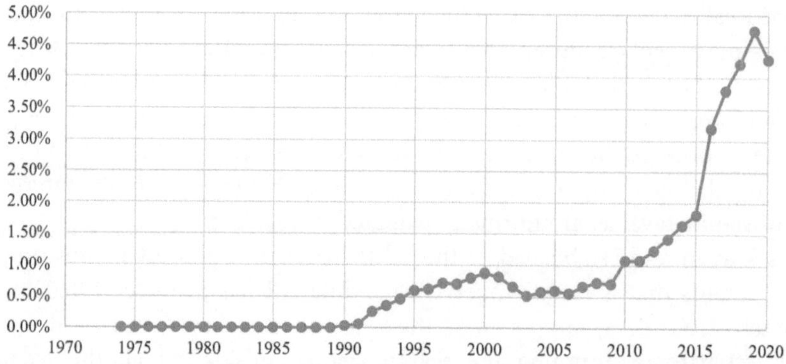

Figure 5. Percentage of metaverse (AR, VR, BCI) patents on all patents. Source: Our elaboration of Espacenet data.

Figure 6 provides information on the distribution of the considered patents by country of the applicant. It indicates a stark predominance of patent applicants from the United States. The U.S. accounts for around 170.000 patents, while the second country - South Korea - accounts for about 52.000 patents, followed by China (33.000) and Japan (23.000).

It is interesting to examine the data suggesting the predominance of U.S. applicants considering a different decomposition of the number of patents, this time by company. Figure 7 presents the 15 companies that hold the largest number of patent portfolios in relation to these technologies. The first place, with more than 8.600 patents, is held by Huawei, followed by LG, with more than 6.000 patents, Microsoft, Tencent, Sony and Samsung follow, all in a range from 4.500 to 4.800 patents, and then Meta, Intel, Oppo and Qualcomm, which each record between 2.950 and 3.300 patents, Google has slightly more than 2.000, and the last group comprises IBM, Apple, Nokia and BOE, who are all in the range between 1.400 and 1.600 patents. The figure, therefore, shows that eight of the fifteen largest patent-holding companies are

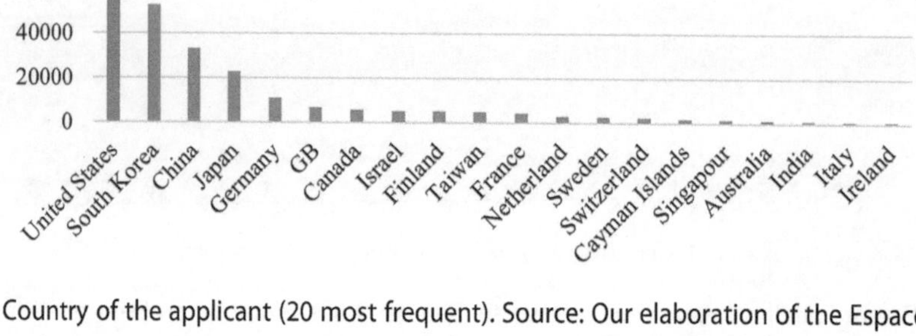

Figure 6. Country of the applicant (20 most frequent). Source: Our elaboration of the Espacenet data.

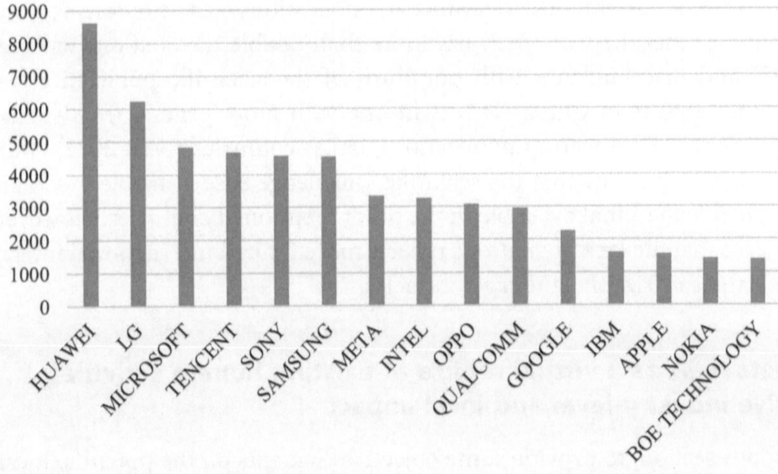

Figure 7. Number of patents per company (top 15). Source: Our elaboration of the Espacenet data.

not American, suggesting that patent endowments of key metaverse-related technologies may be less concentrated in the U.S. than patent applications would imply.

The list of patent holders that emerges from Figure 7, unsurprisingly, includes several innovative technology companies and four of the five GAFAM (Google, Apple, Facebook, Amazon, Microsoft). Whether the standards-based and licensing-based development model typical of the former or the siloed development model of the latter will prevail is currently hard to tell but likely to influence the metaverse's future technological configuration and the extent to which it will be subject to strict control by platforms (on which more below, Section 5).

Additional interesting information may be derived by looking at the countries where metaverse-related scientific publications are produced (Lazzeretti et al. 2022).

Figure 8 shows that the major producers of these publications are China, the U.S. and South Korea. However, a deeper analysis of these results, based on one of the most

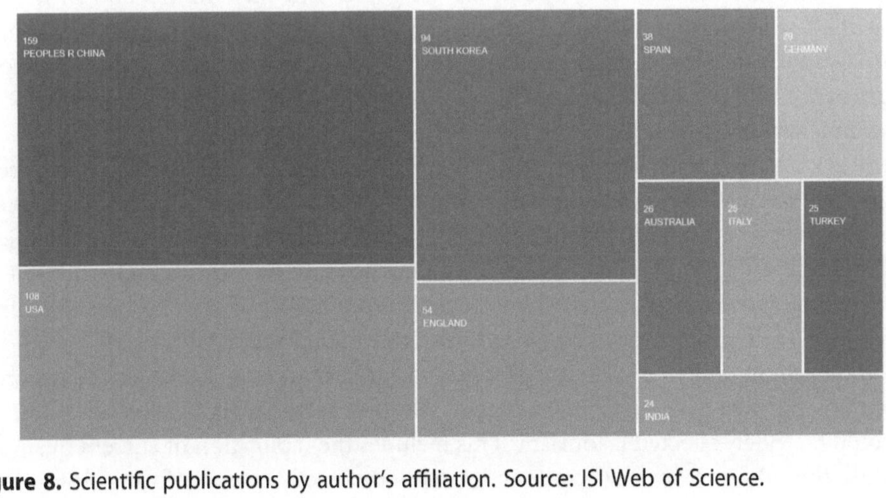

Figure 8. Scientific publications by author's affiliation. Source: ISI Web of Science.

common information to infer the relevance of publications – the citations received – it is possible to notice that the U.S. (682) has more than double the citations with respect to China (297) and that England, with one-third of the scientific publications, received more citations (350) than China. This is in line with more general trends observed in comparing Science&Engineering publications across countries (NSB 2022), which have been interpreted to indicate that the scientific knowledge base available in China is relatively less valuable than that available in the other mentioned countries. However, it may also be the case that the lack of citations reflects more limited international integration of Chinese scientists or simply language obstacles.

4. The metaverse as a virtual replica of existing human activities: prospective industry-level and local impact

In the previous section, to provide some objective elements on the potential local impact of the metaverse, we have considered the production side by looking at the country-level distribution of patents on technologies that may influence the future development of the metaverse and the country distribution of related scientific publications. In this section we broaden the perspective beyond objective measures and enrich our qualitative understanding of the phenomenon, considering the application and use of metaverse technologies at the local level. As mentioned in the introduction, the analysis of this section applies to the 'minimalist' view of the virtual worlds as a collection of technologies that may complement existing human activities. We begin by briefly describing the main sectors affected by the range of technologies that underlie the development of the metaverse. We then highlight three dimensions of these technological developments that we believe will be decisive in influencing their spatial effects, putting forward three hypotheses on their impact and corresponding implications for policy planning.

The most pessimistic view sees the sectoral effects of the metaverse as essentially confined to the gaming sector. In the latter, the virtual worlds-based business model is rather consolidated but what may prospectively make the difference is the potential pervasive use of a group of related instruments (cryptocurrencies, distributed ledger technologies, smart contracts, etc.) that may help developing new business models pushing gamers to increase the time they spend on the game, and their immersion in it, thus raising their willingness to spend. Even in this pessimistic view, it cannot be ignored that innovations developed in gaming may easily spill over to a range of other sectors, and primarily to the traditional content-creation sectors such as the film industry and entertainment more generally.

Early evidence suggests, however, that it is more likely that virtual worlds-related technologies will reach beyond gaming, impacting both the manufacturing and the services sector. Manufacturing is perhaps the sector where the concrete local impact of the metaverse could be most visible. Production lines may be improved by the extended use of 'digital twinning', which allows for the virtual representations of existing business functions, and to realize important savings with extended reality technologies applied to the design and prototyping of products. Also, new forms of marketing and product customization may innovate inventory management. Another sector with a visible local impact that could be widely affected is tourism. This includes the replication of current business, but with the addition of new content through the metaverse platforms, leading to innovative business models like virtual worlds tourism and sustainable virtual worlds tourism (Go and Kang 2022; Koo et al. 2022; Gursoy, Malodia, and Dhir 2022).

Pinning down the local impact of metaverse-related technologies in other services is harder. A common thread appears to be the fact that the potential disruption induced by these technologies may result in a significant wave of innovation and entrepreneurial opportunities. For instance, the virtual worlds can be considered the new frontier for online shopping, which is especially relevant for e-commerce platforms. The sellers may seek the possibility of presenting their physical products in a new, more interactive way to consumers, thus increasing both customers' loyalty and the demand for the products (Jeong, Yi, and Kim 2022).

Real estate, in the virtual worlds, is already a reality, with the possibility of selling finished properties and virtual plots of lands, allowing the user to monetize their investments by either renting or through other forms of management (Nakavachara and Saengchote 2022).

Advertising could also be profoundly impacted. Entering the metaverse gives visibility to small companies (press, blog, social media, etc.) and even more visibility to big brands (Kim 2021; Taylor 2022; Rauschnabel et al. 2022), who have approached the virtual worlds to refresh their image (e.g. Coca-Cola, Gucci, Nike, Luis Vuitton and Balenciaga) and may possibly exploit the NFT market to develop a new profitable line of business for the future (recently Nike acquired on Roblox the digital shoe market RTFKT).

Beyond industry, metaverse-related technologies may transform other domains that may have a significant local impact. Three main areas stand out. The first is risk management, which has a range of different applications that may benefit from virtualization and experimental forms of social interaction. Improvements in risk management may benefit policy planning in many ways, from the city level, where mobility patterns may be better controlled and oriented, to the global scale of climate change problems, which improved predictive abilities may help to address significantly more effectively. Some cities like Shangai, Brisbane and Singapore have already started experimenting with hazard management tools.

Distance education could also be significantly improved by a more immersive experience for students at all stages of study, from elementary school to university specializations (Jeon 2021; Tlili et al. 2022; Hirsh-Pasek et al. 2022).

Another key domain of application of virtual worlds technologies of social relevance is the health sector. VR can create new channels for the delivery of care, potentially lowering costs and improving services to patients, leading to multimodal medical information standards, medical and social data fusion, telemedicine and online health management (Chen and Zhang 2022). Among the tools, telepresence, allowing people to be together virtually even while they are far apart, digital twinning, and blockchain (related to patients' data management and exchange), are expected to become relevant applications.

More generally, public service provision could find interesting new ways of being organized: tapping into metaverse's technology, Seoul Metropolitan Government, for example, is investing significant amounts of money to set an immersive public service platform, which is expected to benefit elder citizens.

In what follows we put forward three hypotheses on the impact of metaverse-related technologies at the local level. First, we argue that the availability of network infrastructures will be a constraint more significant than in the past to the ability of territories to exploit the new technological and entrepreneurship opportunities. Second, the nature of technological opportunities opened up by the evolution of the metaverse will tend to limit the role of agglomeration economies and local knowledge spillovers, provided network infrastructures are locally available. Third, convergence among different sectors will increase, leading to much stronger competitive dynamics that may significantly disrupt existing businesses, leading to significant reallocation of economic activities.

Regarding the first hypothesis, it has been noted in Section 2.2 that the infrastructure layer is the one that currently poses the more significant hurdles to the further development of the metaverse. These hurdles relate to the need for infrastructure to ensure sufficient speed to enable a fully immersive experience, whit response times of around 10 milliseconds. Here it is worth adding that, for this to occur, given the constraints imposed by the basic laws of physics, at least some infrastructure elements should be placed near the end user. Therefore, territories can be active users of metaverse-related technologies only insofar as sufficiently well distributed local infrastructures are available. The speed of rollout of advanced connectivity solutions, such as fibre and 5G networks, has so far been determined by factors that are mostly independent of the development of the metaverse concept. Most countries have adopted policies to accelerate network rollout, but penetration remains strongly uneven at the global level, with the U.S., South Korea, Japan and China standing out in 5G deployment (ETNO 2023). In the European context the DESI Scoreboard (2022), while tracing some overall progress, consistently highlights remarkable differences across Member States (for example in terms of spectrum assignment, fixed broadband subscription, 5G mobile coverage), not to mention those among rural and urban areas. Thus, if these differences remain, it is possible to speculate that the ability of different countries and territories to immediately seize the innovation and entrepreneurial opportunities potentially opened up by the metaverse will be uneven, possibly even to a greater extent than it has previously been the case for the opportunities opened up by Internet 2.0.

Our second hypothesis traces a connection between the nature of the technological opportunities associated with the evolution of the metaverse and the extent of local agglomeration economies. To understand this connection, it is worth starting from the observation that the technologies related to the metaverse are expected to unleash a wave of technological and therefore entrepreneurship opportunities both in manufacturing and in the services sector.[9] In manufacturing, as mentioned, this may prospectively take the shape of the further transformation towards automation of industrial processes, opening the way to the emergence of new firms specializing in the provision of innovative solutions. In the services sector, within the simplified metaverse value chain described in Section 2.2, it is content that promises to have the greatest local impact in terms of creation of entrepreneurial opportunities, change in consumers/users' behaviours, and market disruption. Both the nature of the economic activities that (may/will) potentially belong to the *Content layer* and its prospective economic impact may go much beyond what is currently referred to as 'content' in Internet 2.0. This can be immediately perceived by observing, referring to the technological illustration elaborated by a prominent industry expert (Radoff 2022), that the *Content layer* can be said to encompass three sublayers: *Creator Economy*, *Discovery* and *Experience*. The expression *Creator Economy* refers to all the technologies that content creators use to craft the experiences that people may enjoy in the virtual worlds, i.e. the combination of software and marketplaces that make it possible for creative people and teams to fill the virtual spaces. Companies like Roblox, Decentraland, Epic Games, ones that are either completely new or that come from the game industry, compete with the more well-known digital players, like Microsoft, Adobe, Shopify.

The domain named *Discovery* constitutes a vast ecosystem, presently one of the most lucrative for companies, that includes all the technologies through which, directly or indirectly, information and experiences that can be carried out within the virtual worlds are presented and offered to users. In addition to search engines, curation, and advertisement, in the virtual worlds it is expected that, given the core value of social interaction, community-driven discovery will gain relevance. In this area, Big Techs, with a consolidated and powerful advertising business model, from Google to Facebook, will directly compete with the new inventors of content, from gamers to space creators, like Unity and Unreal. Nonetheless, in the future, community-driven discovery that exploits real-time presence may lead the way if, as expected, the virtual worlds foster the transition from asynchronous social networking to real-time and collective virtual social activities.

Finally, the domain named *Experience* is the content layer in a more properly traditional sense. Here, the competition is open to all content producers that aim at a market that is constituted by the attention and the time of the metaversers. This is the realm for entertainment companies, from Netflix to YouTube, from connection companies like Zoom to social media like Facebook, and today is primarily the domain of the gamer companies as Fortnite, Activision Blizzard, Minecraft, etc.

Increased entrepreneurship opportunities arise in all three layers, although the Discovery domain is certainly characterized by higher barriers to entry. Innovation and entrepreneurship in these domains aim at participating in global markets and require creative ideas and mostly top-down codified technical knowledge that is accessible through global epistemic communities rather than bottom-up cumulative knowledge. This entails that the stock of existing localized knowledge assets is not crucial to seize the technological opportunities offered by the virtual worlds, so that new patterns of local development may emerge with the potential to escape to traditional path dependence constraints. Thus, technological opportunities in the creation of virtual products and experiences may possibly limit the role of agglomeration economies and local knowledge spillovers, which the advent of the intangible economy has, somewhat surprisingly, even reinforced (Haskel and Westlake 2018).

The strength of this hypothesis may also be buttressed by the observation that a technological trend exists in the content layer of the metaverse towards the decoupling of good business ideas from coding capabilities and technical competence based on the increasing availability of ready-to-use software tools accessible to any would-be entrepreneur. Indeed, since the success of the emerging virtual worlds is linked to the ability to spur direct and indirect network effects so as reach a critical mass of users, existing companies active in the Creator Economy domain have an interest in lowering the barriers to participation for any user-developer that may help to build the content needed to engage other users in an immersive experience. If the trend continues, the diffusion of metaverse-related technologies may also improve social inclusion, by alleviating some of the constraints to entrepreneurship. This would be particularly likely if the most idealistic views of Decentralized Finance (DeFi) linked to the evolution of Web 3.0 did effectively contribute to alleviate also financial constraints to entrepreneurial activities (OECD 2022).[10]

It is important to stress, however, that the second hypothesis, as well as the third, to be put forward next, may hold provided that infrastructures are homogeneously available across territories. The previous (and the following) discussion indicate that local conditions may constitute particularly strong constraints to entrepreneurship. Forms of digital divide, with more extreme implications than in the traditional internet, and unequal distribution of innovation opportunities across territories may indeed be the outcome of persisting inequalities in the distribution of infrastructures.

The third hypothesis on the local effect of the metaverse relates to the degree of convergence among sectors that the metaverse may determine, which is substantially magnified with respect to the traditional Internet. Convergence has been a disruptive force in Internet 2.0, but limited mostly to digital(ized) economic activities or to vertical relationships among sectors. With voice, data and broadcasting simply becoming a stream of packages, competition has intensified across the layers of the traditional value chains, for instance in broadcasting, where firms from traditionally distinct rings of the chain have started competing. In the most sanguine views of the metaverse, by contrast, horizontal convergence across sectors appears likely. For instance, innovations from the gaming industry may be used across a range of different sectors, giving rise to new competitive dynamics. Moreover, convergence will not only be confined to the digital world, but it may prospectively become a matter of physical-digital integrations, giving rise to unexpected innovation opportunities.

Two main implications of the increased degree of convergence can be traced for local economies. The first is that, as it has been the case for previous generations of the Internet, increased convergence entails greater competition and therefore a significant disruption of existing economic activities and business models. A significant wave of reallocation of activities and factors of production may be expected to follow, which may create social tensions and inefficiencies due to reallocation frictions. The second implication of convergence is that it may favour evolutionary scenarios whereby Big Tech platforms, by exploiting their simultaneous presence on multiple convergent domains, play an ever-increasing role in mediating access to global markets by local entrepreneurs and therefore exert an increasing influence on local development. This possible evolution is further explored in the following section.

In the instantiation of the metaverse considered in this section, the main challenge for planning concerns understanding the changes the 'metaverse' may potentially trigger in traditional industries and economic activities, which in turn feed back into changes in the local economies and adjust spatial development policies accordingly. The analysis proposed suggests that the main concerns for planning are not substantially different from previous instances of disruptive technological change, albeit possibly on a grander scale. More specifically, the significant wave of innovation opportunities that the metaverse may bring about calls for appropriate policies to avoid missing opportunities. Our first and second hypothesis together indicate that a precondition of this would be policies aimed at ensuring the homogeneous availability of appropriate network infrastructures across territories. This provides a rationale for public support to the rollout of advanced network solutions that allows to overcome resistance due to acknowledgement of the present lack of user demand for such infrastructures. The second hypothesis suggests the need for digital industrial policies targeting both technology adoption and local innovation capabilities (Gruber 2019). Since we believe the ability of local territories to escape path dependent patterns of development will be partly linked to the ability to seize the entrepreneurial opportunities associated with metaverse-related technologies, we advocate policies focused on the provision of the required basic and advanced digital skills.

Finally, our third hypothesis suggests that policies mitigating the effects of the reallocation of economic activities that may follow from increased convergence are certainly needed. More generally, a major role for planning would be linked to the need to address the likely employment consequences of this variant of disruptive technological change. It is too early to tell whether job-creation or job-destruction effects from innovation in these technological areas will prevail, but there is no doubt that policies capable of promoting the efficient reallocation of labour across different sectors and industries, as well as to mitigate the social effects of the transition, will be key to prevent undesirable local impacts.

5. The metaverse as an entirely new virtual space: key dimensions of technological and market evolution

A more extreme realization of the metaverse sees it as an entirely new environment for social interaction that, although virtual, may change more profoundly the economy by altering certain basic institutions: *scarcity*, which may become the artificial product of software designers' choices rather than the outcome of objective resource constraints; *property*, whose scope may be entirely redefined by distributed ledger technologies; and *markets*, which may expand and become much more granular. Many more intermediate versions of the metaverse may become reality. In what follows, we explore three key dimensions of development that we believe will play a role in influencing the evolution of the metaverse and sketch some possible associated policy implications. Differently from the previous section, due to the higher degree of uncertainty that surrounds this speculative scenario, we do not formulate specific hypotheses and we limit ourselves to indicating possible evolutionary paths. In any event, at stake is an entirely new role for policy planning, one that involves directing the evolution of the new virtual worlds in desired directions and considers the feedbacks on the real world. The three dimensions have been selected because they reflect the main topics of current debate. They are: the degree of interoperability, of platformization/centralization, and of financialization of the virtual world(s), as better explained below.

Let us start from the degree of interoperability: will there be only one, or will there be a plurality of metaverses, which are only partially able to communicate? The World Economic Forum has recently clarified that, in the metaverse, 'interoperability depends on data interchange across infrastructures to enable participants' ability to access, move, transact and create within and across digital (and physical) worlds' (WEF 2023, 6). This entails making choices at the technical level, related to the ability to connect different networks and technologies pertinent to users' virtual experiences, but also choices in terms of rights and forms of use of metaverse-related technologies, as well as choices related to the ability to guarantee the same type and level of metaverse experience across different geographies and jurisdictions. Standards of interoperability would define these choices.

A first issue for policy planning in this connection thus concerns identifying the desirable degree of interoperability and the consequent need for standardization. Clearly, greater interoperability is associated to greater user value in terms of direct and indirect network effects: users' benefits from the metaverse experience increase with the ability to enjoy multiple immersive experiences seamlessly and with the increased availability of tools and solutions brought about by the standardization of development tools. However, users' wellbeing may potentially be negatively affected if more interoperability implies less privacy, security or safety. Ascertaining the extent of these trade-offs, which are difficult to describe, let alone quantify today, appears difficult for traditional public policy making due primarily to lack of the relevant information.

This raises a second important issue: who should develop interoperability standards? There is clearly a role for public policy to ensure that design choices reflect basic human rights and are not excessively conditioned by different stakeholders' private interests. Indeed, one key feature of the metaverse as a new immersive reality is that it will be the outcome of designers' choices: the entire ecosystem would be imagined and programmed by designers and humans would interact with machines that are programmed by other humans. This suggests that desirable features reflecting public policy objectives, such as for instance inclusion and non-discrimination, should be embedded from the start *through standards* in metaverse design. However, there is clearly the need to leverage on private stakeholders' information to address the knowledge gap of public decision-makers and thus a need for effective public-private collaborations.

There is also a geographic dimension of the policy issue, as different jurisdictions will want to embed their own vision for public policy values in standardization efforts, as it is already occurring in the realm of artificial intelligence. This presents a further trade-off for local planning. The emergence of this vision of the metaverse is likely to determine a further weakening of the link between local effects and local decision-making (possibly much beyond what has happened through globalization). The nature of users' experience in the metaverse will be determined by choices made far from the local level. Thus, national and local planners may have to decide whether to participate to the full metaverse experience, as it will emerge from public- and private-led design choices taken at the supra-national level, or enforce local rules meant to protect local cultural norms and values, at the cost of a reduced ability to partake to the metaverse experience.

The second dimension to which we draw attention is platformization, i.e. the extent to which digital interactions in the metaverse will be intermediated by platforms, and particularly by the old GAFAM.

Two possible scenarios that pose different policy challenges may emerge, together, of course, with a multitude of intermediate solutions. In the first, current Big Tech platforms, starting from Meta, succeed in orienting the evolution of the metaverse in a way that preserves their central role in digital interactions and expands it to virtual interactions. The alternative is the vision of the Internet that goes under the name of Web 3.0, i.e. Internet as a collection of decentralized applications built on distributed blockchain ledgers (Marchetti 2022). The Web 3.0 concept synthesizes a potential paradigm shift aimed primarily at reducing the weight of centralized servers and platforms in the Internet architecture and restituting control, particularly over their personal data, to users.

Distributed ledger technologies could provide an alternative to platforms to solve the problem of trust at the heart of digital interactions. Such evolution would reduce the power of intermediaries and gatekeepers facilitating the emergence of new decentralized organizational forms.

As for the first scenario, the five major Big Techs, the old GAFAM, are active in every layer of the metaverse value chain, and particularly in what we have described as the Discovery sub-layer. However, they are apparently following different strategies to face the challenge that Internet 3.0 and the metaverse pose to their specific areas of dominance in Internet 2.0. In the previous classification of layers, Meta is mainly located in the Human Interface (Oculus), in the Discovery and in the Creator Economy (Horizon World). Apple has recently declared it is betting on AR (in which virtual elements and images are superimposed onto the real world), thus attempting to confirm its primary role as a producer of sophisticated devices, but is also active in Discovery. Microsoft is probably one of the few companies that seems to be investing in internal R&D transversally in all the possible layers and that, at the same time, is attempting to escalate its presence with acquisitions (e.g. Activision Blizzard in gaming and AltspaceVR in social networking). Somehow in between are Amazon and Google, whose plans for the metaverse have not yet been openly explained. They both seem to be investing mainly internally and especially in the infrastructure layer (mostly through their cloud business), but it is known that they have many other ongoing technological projects (for example Google AI in spatial computing).

A scenario whereby current Big Tech platforms succeed in transferring their leverage over social interactions to the virtual world(s) is one that would justify policy responses aimed at limiting platforms' concentration of power to an even more radical extent than is presently the case. This is an evolution that has already been highlighted in the context of the debate over the concept of 'smart cities' (Cooke 2021). In such a scenario, Big Techs would be able to embed in their technological choices their preferred elements of the design of social interaction, thus shaping the nature of the experiences possible in the new virtual spaces largely outside of any public oversight, going even beyond Lawrence Lessig's perspective that 'code is law' (Lessig 2000). How likely is it that the most salient issue for policy planning will be addressing platforms' concentration of power in shaping the metaverse? This is clearly hard to tell today. Platforms will certainly play a role in making the metaverse as much 'user-friendly' as possible and therefore in facilitating adoption. This has been their key role in Internet 2.0, where they have acted as intermediaries sustaining trust and making accessible direct and indirect network effects from users' interaction. However, there may be limits to the trust that users may put in Big Techs when the latter have the last word on every aspect of the metaverse experience. In a platform-dominated scenario, the value of any piece of virtual land, the value of any transaction, and even the very existence of any avatar in the metaverse would be subject to the *fiat* power of the relevant platform. In economic terms, users' investments in any activity on the metaverse would thus become extremely specific and exposed to risks of hold-up by the platform. As with any other kind of specific investment, this suggests that there would be a problem of users' underinvestment in activities performed on a platform-dominated metaverse. In other words, while it is highly probable that platforms will play some intermediary role in the metaverse, there may also be endogenous forces pushing in the opposite direction.

Another force pushing in the opposite direction, partly technological and partly ideological, is Web 3.0 decentralization. Key components of this evolution would be new 'institutions' like Decentralized Applications (DApps) and Decentralized Autonomous Organizations (DAOs). Both rely on collaboration among Internet users, sustained through a system of well-defined embedded incentives based on the possibility to monetize actions and data. The first (DApps) refers to the new type of services that can be provided based on smart contracts and tokens. The second (DAOs) refers to the main business model that is currently a candidate to be a substitute for Big Techs in Web 3.0: collaborative organizations set up in the pursuit of a common (business, ethical or social) goal and managed through an open, transparent and participatory governance system (Goldberg and Schär 2023).

Essential to the emergence of this vision for the future would be a much greater *degree of financialization* of digital interactions. The alignment of individual and collective incentives would rely on the possibility for users to monetize their actions and data by obtaining tokens and on crypto-based payments. Tokenization, in turn, may provide incentives for users' participation in the metaverse(s), thus alleviating the demand-side constraints that are currently hampering its emergence. This also means that each digital object would become clearly identifiable, with an expansion of the realm of property rights that goes much beyond what legal coding has done in the intangible economy by turning any object, promise or idea into a capital asset (Pistor 2018).

In this scenario, the challenge for public policy would be more extreme. Indeed, at the core of the vision behind decentralized technologies is the idea that systems of incentives based on game-theoretical principles aimed at aligning private interest and the public good and mechanism design can avoid not only centralized control by current intermediaries but also centralized control by governments. According to Gavin Wood, who has coined the term, Web 3.0 is about building systems that do not rely on trust in people, corporations or governments (Wood 2018). In other words, technology design should be considered a substitute for public policy design. As per this view, for instance, distributed ledger-based decentralization would be a substitute for public regulation in the form of data portability and interoperability measures.

In the (unlikely) event that it fully materializes, this view may potentially alter the balance between public and private power even more deeply than it has been the case with the emergence of platforms in Internet 2.0. Governance rules and incentive systems defined within specific DAOs would define rules of behaviour in a multitude of new virtual environments, on which it would in any case be harder to exercise any public oversight. This is, of course, an extreme and possibly unlikely evolution of the virtual space. However, it underscores the case for technology-aware advanced planning of policy responses, to ensure coherence between technological solutions and public policy objectives. One evident example and an area that is of immediate relevance in this connection is the regulation of cryptocurrencies, which are currently governed by companies that are hard to make accountable (Marchetti 2022). The present cryptocurrency crisis and the associated risk of contagion suggests that we are already grappling with these issues. This very example suggests that design's failures in the virtual world may exercise relevant negative feedbacks also on the real world and it is the responsibility of foresighted public policies to anticipate and avoid these consequences.

5. Conclusions

How real will the metaverse be? It is now clear that the initial ambitions with which the concept has entered the public discourse and has attracted attention in the scientific and entrepreneurial communities have rapidly scaled back. In this paper we have attempted to go beyond the recent hype and trace two possible evolutive scenarios. In the first, the metaverse will remain nothing more than a fashionable label that technology companies in a range of frontier domains (AR, VR, BCI and others) may invoke to make their products more appealing. In this scenario, applications will be limited to replicating existing human activities in an immersive virtual environment and it will be physical reality to drive and shape the metaverse. In the second, more imaginative, scenario the metaverse will evolve in a (set of) virtual worlds that will enable new forms of social interactions, substantially affecting key economic concepts and institutions.

In both cases, there will be concrete economic effects spilling over from the virtual to the real environment and there is a need to adapt planning, although the nature of the policy challenges raised will be different. In the more 'minimalist' and disenchanted view of the metaverse, the core policy challenge will be traditional, albeit possibly on a larger scale. Attention should be primarily directed towards policies capable of enhancing local abilities to exploit the innovation opportunities that frontier technologies open up in a wide range of sectors, to prevent the emergence of excessive territorial inequalities and eventually to accompany the reallocation of labour necessitated by disruptive technological change and to mitigate its social effects. If, by contrast, the concept of virtual worlds gains viability and attracts a critical mass of users, the innovative challenge of planning policies for the new virtual space will emerge. In this connection, we have highlighted three key dimensions of evolution of the metaverse(s) - degree of interoperability, platformization, and financialization - identifying some of the policy challenges they may prospectively entail. Planning will be needed to ensure that desirable features are incorporated from the start in the framework within which new forms of social interaction will take place. The article has only sketched a few elements of the reasoning required to start making sense of the spatial impact of the metaverse, with many limitations due also to the inevitably speculative nature of the qualitative analysis proposed. Nonetheless, we hope to have pointed to relevant areas for further research. First, a more refined quantitative mapping of the technologies underlying the production of metaverse components as well as their adoption from a regional perspective is in order, also exploring, for example, the distance of Europe from the technological frontier of the development of key technologies. Second, there is a need for a more precise analysis of the correspondence between metaverse-related technologies and the digital skills and resources indispensable at the local level to capture the opportunities they open up, so as to inform policy planning. Relatedly, it is important to explore the potential employment impact of AR and VR, as it cannot be taken for granted that the conclusions presently reached for AI can directly apply also to these technologies. Moreover, additional research will need to refine our understanding of the relationship between the degree of interoperability, platformization and financialization of the virtual words and local economic outcomes. Finally, it will be important to explore whether and how the emergence of viable virtual worlds may open an important opportunity for planning by providing useful sandboxes for policy experimentation that may improve the way planning is defined and implemented.

Aknowledgements

The authors would like to thank the editors of this journal, two anonymous referees as well as the participants and speakers at the research seminar 'Strolling in the Metaverse', held at the European University Institute on 14 December 2022. The usual disclaimer applies.

Notes

1. Park and Kim (2022) report 54 different definitions of the metaverse.
2. See the following article on the state of the art of the ongoing reviews of the merger https://www.polygon.com/23546288/microsoft-activision-blizzard-acquisition-deal-merger-ftc-latest-news
3. ISI Web of Science search based on the word 'metaverse' using topic as search criteria in all scientific fields (6 December 2022).
4. The hype in the scientific community precedes that of the public. Figure 2 shows ISI publications on scientific journals. Therefore, papers published in 2022 have likely undergone peer-review processes and were prepared in the previous years.
5. https://www.roundhillinvestments.com/etf/metv/.
6. While agreed-upon definitions are hard to provide, it is widely accepted that Internet 2.0 refers to the era of the dynamic Internet (as opposed to the static Internet 1.0), focused on reading and writing content, while Internet 3.0 will focus more on content creation and as explained in Section 5, may possibly be more distributed.
7. However, this skewed age profile can also be interpreted as being good news for the future of the metaverse concept.
8. There are devices at lower prices, some with a price lower than $100 (e.g. the Sony PlayStation VR), these are commonly used only for gaming, leaving few possibilities for those who are interested in the non-gaming metaverse(s).
9. We use this distinction for ease of exposition, but we acknowledge that the distinctions between services and manufacturing are ever more blurred by digitalization.
10. The expression DeFi refers to financial applications built and run on a blockchain, enabling smart contracts-based virtual transactions with no central authorities or intermediaries.

Disclosure statement

No potential conflict of interest was reported by the author(s).

ORCID

Pier Luigi Parcu http://orcid.org/0000-0003-2156-7308
Maria Alessandra Rossi http://orcid.org/0000-0003-2480-7266
Niccolò Innocenti http://orcid.org/0000-0001-8421-5479
Chiara Carrozza http://orcid.org/0000-0002-3402-9990

References

Acs, Z., L. Anselin, and A. Varga. 2002. "Patents and Innovation Counts as Measures of Regional Production of new Knowledge." *Research Policy* 31(7): 1069–1085. doi:10.1016/S0048-7333(01)00184-6

Anderson, P., and M. Tushman. 1990. "Technological Discontinuities and Dominant Designs: A Cyclical Model of Technological Change." *Administrative Science Quarterly*, 604–633. doi:10.2307/2393511

Atkinson, A. 2015. *Inequality: What Can Be Done?* Cambridge, MA: Harvard University Press.

Ball, M. 2022. *The Metaverse: And how it Will Revolutionize Everything*. New York, NY: Liveright Publishing.

Belloc, F., and U. Pagano. 2012. "Knowledge Enclosures, Forced Specializations and Investment Crisis." *The European Journal of Comparative Economics* 9 (3): 446–483.

Burhan, M., A. Singh, and S. Jain. 2017. "Patents as Proxy for Measuring Innovations: A Case of Changing Patent Filing Behavior in Indian Public Funded Research Organizations." *Technological Forecasting and Social Change* 123: 181–190. doi:10.1016/j.techfore.2016.04.002

Chen, D., and R. Zhang. 2022. *Exploring Research Trends of Emerging Technologies in a Health Metaverse: A Bibliometric Analysis*. Available at SSRN 3998068.

Cooke, P. 2021. "After the Contagion. Ghost City Centres: Closed "Smart" or Open Greener?" *Sustainability* 13 (6): 3071. doi:10.3390/su13063071

Dernis, H., M. Squicciarini, and R. de Pinho. 2015. "Detecting the Emergence of Technologies and the Evolution and co-Development Trajectories in Science." *Journal of Technology Transfer* 41: 930–960. doi:10.1007/s10961-015-9449-0

Digital Economy and Society Index (DESI) Report. 2022. "Connectivity." https://ec.europa.eu/newsroom/dae/redirection/document/88766

Dwivedi, Y., L. Hughes, A. Baabdullah, S. Ribeiro-Navarrete, M. Giannakis, M. Al-Debei, and S. Wamba. 2022. "Metaverse Beyond the Hype: Multidisciplinary Perspectives on Emerging Challenges, Opportunities, and Agenda for Research, Practice and Policy." *International Journal of Information Management* 66: 102542. doi:10.1016/j.ijinfomgt.2022.102542

ETNO (European Telecommunications Network Operators' Association). 2023. The State of Digital Communications 2023.

Financial Times. 2023. "Meta Axes Further 10,000 Jobs in Fresh Round of Cuts.".

Go, H., and M. Kang. 2022. "Metaverse Tourism for Sustainable Tourism Development: Tourism Agenda 2030." *Tourism Review* 78 (2): 381–394. doi:10.1108/TR-02-2022-0102

Goldberg, M., and F. Schär. 2023. "Metaverse Governance: An Empirical Analysis of Voting Within Decentralized Autonomous Organizations." *Journal of Business Research* 160: 113764. doi:10.1016/j.jbusres.2023.113764

Gruber, H. 2019. "Proposals for a Digital Industrial Policy for Europe." *Telecommunications Policy* 43 (2): 116–127. doi:10.1016/j.telpol.2018.06.003

Gursoy, D., S. Malodia, and A. Dhir. 2022. "The Metaverse in the Hospitality and Tourism Industry: An Overview of Current Trends and Future Research Directions." *Journal of Hospitality Marketing & Management* 31 (5): 527–534. doi:10.1080/19368623.2022.2072504

Haskel, J., and S. Westlake. 2018. *Capitalism Without Capital: The Rise of the Intangible Economy*. Princeton: Princeton University Press.

Hirsh-Pasek, K., J. Zosh, H. Hadani, R. Golinkoff, K. Clark, C. Donohue, and E. Wartella. 2022. "A Whole New World: Education Meets the Metaverse." *Policy Brief*, February 2022. Brookings Institution.

Jeon, J. 2021. "A Study on Education Utilizing the Metaverse for Effective Communication in a Convergence Subject." *International Journal of Internet, Broadcasting and Communication* 13 (4): 129–134.

Jeong, H., Y. Yi, and D. Kim. 2022. "An Innovative e-Commerce Platform Incorporating the Metaverse to Live Commerce." *International Journal of Innovative Computing, Information and Control* 18 (1): 221–229. doi:10.24507/ijicic.18.01.221

Kim, J. 2021. "Advertising in the Metaverse: Research Agenda." *Journal of Interactive Advertising* 21 (3): 141–144. doi:10.1080/15252019.2021.2001273

Koo, C., J. Kwon, N. Chung, and J. Kim. 2022. "Metaverse Tourism: Conceptual Framework and Research Propositions." *Current Issues in Tourism*, 1–7. doi:10.1080/13683500.2022.2122781

Kraus, S., D. Kanbach, P. Krysta, M. Steinhoff, and N. Tomini. 2022. "Facebook and the Creation of the Metaverse: Radical Business Model Innovation or Incremental Transformation?"

International Journal of Entrepreneurial Behavior & Research 28 (9): 52–77. doi:10.1108/IJEBR-12-2021-0984

Lazzeretti, L. 2023. *The Rise of Algorithmic Society and the Strategic Role of Arts and Culture.* Cheltenham.: Edward Elgar.

Lazzeretti, L., N. Innocenti, M. Nannelli, and S. Oliva. 2022. "The Emergence of Artificial Intelligence in the Regional Sciences: A Literature Review." *European Planning Studies*, 1–21. doi:10.1080/09654313.2022.2101880

Lessig, L. 2000. *Code and Other Laws of Cyberspace.* New York, NY: Basic Books.

Marchetti, S. 2022. "Web3, blocksplained. Questioni di Economia e Finanza (Occasional Papers) 717." Bank of Italy, Economic Research and International Relations Area.

Metaversed. 2022. *Metaverse User age Profile. Metaversed Research & Insight.*

Mozumder, M., M. Sheeraz, A. Athar, S. Aich, and H. Kim. 2022. "Overview: Technology Roadmap of the Future Trend of Metaverse based on IoT, Blockchain, AI Technique, and Medical Domain Metaverse Activity." Paper presented at the 24th International Conference on Advanced Communication Technology (ICACT), 256–261.

Nakavachara, V., and K. Saengchote. 2022. "Is Metaverse LAND a Good Investment? It depends on Your Unit of Account!" *arXiv* preprint arXiv:2202.03081.

OECD. 2022. *Why Decentralised Finance (DeFi) Matters and the Policy Implications.* Paris: OECD.

Parcu, P., N. Innocenti, and C. Carrozza. 2022. "Ubiquitous Technologies and 5G Development. Who is Leading the Race?" *Telecommunications Policy* 46 (4): 102277. doi:10.1016/j.telpol.2021.102277

Park, S., and Y. Kim. 2022. "A Metaverse: Taxonomy, Components, Applications, and Open Challenges." *IEEE Access* 10: 4209–4251. doi:10.1109/ACCESS.2021.3140175

Pistor, K. 2018. *The Code of Capital.* Princeton: Princeton University Press.

Radoff, J. 2022. Building the Metaverse.

Rauschnabel, P., B. Babin, M. Dieck, N. Krey, and T. Jung. 2022. "What is Augmented Reality Marketing? Its Definition, Complexity, and Future." *Journal of Business Research* 142: 1140–1150. doi:10.1016/j.jbusres.2021.12.084

Taylor, C. 2022. "Research on Advertising in the Metaverse: A Call to Action." *International Journal of Advertising* 41 (3): 383–384. doi:10.1080/02650487.2022.2058786

Tlili, A., R. Huang, B. Shehata, D. Liu, J. Zhao, A. Metwally, and D. Burgos. 2022. "Is Metaverse in Education a Blessing or a Curse: A Combined Content and Bibliometric Analysis." *Smart Learning Environments* 9 (1): 1–31. doi:10.1186/s40561-022-00205-x

Vidal-Tomás, D. 2022. "The new Crypto Niche: NFTs, Play-to-Earn, and Metaverse Tokens." *Finance Research Letters*, 102742. doi:10.1016/j.frl.2022.102742

Vidal-Tomás, D. 2023. "The Illusion of the Metaverse and Meta-Economy." *International Review of Financial Analysis* 86: 102560. doi:10.1016/j.irfa.2023.102560

Wood, G. 2018. Why We need Web 3.0. Medium.

World Economic Forum. 2023. Interoperability in the Metaverse.

Index

Note: Figures are indicated by *italics*. Endnotes are indicated by the page number followed by 'n' and the endnote number e.g., 20n1 refers to endnote 1 on page 20.